Improving Testing

Applying Process Tools and Techniques to Assure Quality

To Lynn Olson —
Thanks you for your coverage of
education and testing! Public awareness
is critical to improving testing.

Cheryl

Improving Testing

Applying Process Tools and Techniques to Assure Quality

Edited by

Cheryl L. Wild

Rohit Ramaswamy

 Lawrence Erlbaum Associates
Taylor & Francis Group

New York   London

Lawrence Erlbaum Associates
Taylor & Francis Group
270 Madison Avenue
New York, NY 10016

Lawrence Erlbaum Associates
Taylor & Francis Group
2 Park Square
Milton Park, Abingdon
Oxon OX14 4RN

Printed in the United States of America on acid-free paper
10 9 8 7 6 5 4 3 2 1

International Standard Book Number-13: 978-0-8058-6469-4 (Softcover) 978-0-8058-5896-9 (Hardcover)

Library of Congress Cataloging-in-Publication Data

Wild, Cheryl L.
    Improving testing : applying process tools and techniques to assure quality / Cheryl L. Wild and Rohit Ramaswamy.
        p. cm.
    Includes bibliographical references and index.
    ISBN 978-0-8058-5896-9 (hardback : alk. paper) -- ISBN 978-0-8058-6469-4 (pbk. : alk. paper)
    1. Testing industry--Management. 2. Process control. 3. Quality assurance. I. Ramaswamy, Rohit, 1960- II. Title.

HD9999.T452W55 2007
153.9'4000684--dc22   2007017205

Visit the Taylor & Francis Web site at
http://www.taylorandfrancis.com

Contents

Foreword

Over 15 years ago the National Commission on Testing and Public Policy (1990, p. 21) stated, "Today those who take and use many tests have less consumer protection than those who buy a toy, a toaster, or a plane ticket. Rarely is an important test or its use subject to formal, systematic independent professional scrutiny or audit." This assertion continues to be true.

The recent publicity in public newspapers regarding testing errors associated with high-stakes examinations has stimulated new debate about the quality controls in place within testing agencies that are responsible for giving these type of tests. Unfortunately, these test errors have affected many individuals by limiting and/or denying them educational or work opportunities.

Although the previously mentioned problem areas were identified many years ago and in a recent report, there is no movement to develop an independent program to recognize or accredit agencies in the testing industry against a consensus standard accepted by the community of interest. This lack of oversight has also been noted in the *Education Sector Report of 2006* titled "Margins of Error: The Education Testing Industry in the No Child Left Behind Era" (Toch, 2006) where there is a recommendation for both state and federal policymakers to strengthen the nation's testing infrastructure. In spite of these reports, no agency such as a Federal Aviation Administration or Council on Higher Education Accreditation has been created to provide oversight and monitoring in this particular industry.

In 2003, the American National Standards Institute, a nonprofit 501C3 organization, began an accreditation program for personnel certification agencies (which involves some of the testing agencies) under the ANSI/ISO/IEC 17024 Standard—*General Requirements for Bodies Operating Certification of Persons*. It is the only accrediting agency in the United States that requires personnel certification agencies and their testing agencies to have a management system in place in order to be accredited. This requirement is designed to ensure an agency has a continuous quality improvement mechanism in place to monitor and identify testing errors in a systematic manner and take timely preventive and/or corrective actions.

Although consumers and government have called for more accountability of testing agencies, there appears to be confusion among interested stakeholders as to how this increased accountability can be achieved. This book provides one of the ways the testing industry could start becoming more accountable by using some of the tools from the quality world. If the strategies and concepts described in this book were formally adopted by the testing industry and known to the public, the consumers of tests could feel more assured that the human and/or nonhuman errors in testing would be reduced. This book provides a variety of models and examples for achieving more accountability through a continuous quality improvement focus that ultimately concentrates on customer satisfaction.

Roy A. Swift, PhD

Program Director
American National Standards Institute

References

National Commission on Testing and Public Policy. (1990). *From gatekeeper to gateway: Transforming testing in America*. Chestnut Hill, MA: National Commission on Testing and Public Policy.

Toch, T. (2006). Margins of error: The education testing industry in the no child left behind era. Washington, DC: Education Sector Report.

Preface

Improving Testing: Applying Process Tools and Techniques to Assure Quality demonstrates how tried and proven tools used to achieve process quality in manufacturing and service organizations can be applied to the testing industry. These tools and methods have been part of standard operations in a variety of industries for the past 20 years, yet transfer of knowledge to testing has been slow. Test sponsors, developers, and vendors view themselves as being more aligned to education, research, and development organizations rather than to the manufacturing and service industries. Sponsors of test programs are often not-for-profit organizations or governmental organizations that have been less demanding of supplier quality than customers of commercial products and services. However, as competition has increased and customer tolerance for mistakes has diminished over the past few years, leading organizations in the testing field have come to realize that process improvement methods are critical to the ongoing success of their business. A few have begun to adopt these methods to improve the quality of their processes.

In this book, best practice organizations in the testing community have contributed chapters and examples to illustrate their use of process-oriented tools and methodology to improve quality. The purpose behind this book is to educate testing organizations (whether vendors, sponsors, or users of tests) about these tools and to encourage them to adapt and adopt these tools to significantly increase the satisfaction of all stakeholders with testing products and services.

The need for a systematic and in-depth approach to improving testing process is great. Educational testing in the United States has more that doubled as a result of the No Child Left Behind legislation. Internationally, certification testing is booming. More and more, organizations are using testing of new hires as a supplement to traditional interviews, recommendations, and resumes. Along with the increased testing, and maybe because of it, has come a plethora of news stories of problems with testing.

Such stories tend to impinge the entire testing industry. News about a scoring issue by one contractor for one test casts a dark shadow on testing in general, however unfair that may be to testing programs that operate according to high standards and well-considered procedures. Of course, some negative publicity comes from small problems blown out of proportion by those who don't believe in testing and wish it would go away. But many articles sound justified alarms about incorrectly scanned answer sheets, essay tests scored according to inconsistent criteria, lost answer sheets, misapplied answer keys, poorly trained test administrators, late test books, test books with errors in them, late test score reports, entire item banks from testing programs for sale on the Web—the list goes on and on. Clearly, any of these errors can have sobering consequences to test takers—artificially low scores may prevent students from graduating from high school, prevent students from being accepted by or even applying to the college of their choice, or prevent adults from working at the job of their choice because of a denied or delayed certification or licensure.

The need to improve testing is thus recognized by most of the testing community. However, the most response of the testing community to this need has historically been to develop new standards related to the testing product, for example, developing standards for computer based testing or for adapting tests to different cultures or translation tests. Although these psychometric, product-related concerns are necessary, they are not sufficient to address the procedural concerns that allow a test to be scanned improperly or any of the other mundane mishaps that can cause glitches in the result of testing.

Improving Testing is the first book on the testing market that focuses on the use of business process tools to improve the quality and the validity of tests. Although there are hundreds of books about quality tools, most use examples from manufacturing and more recently from service industries. The testing arena has received little if any attention in the quality field. What makes this book unique is that the authors have all worked with testing in some way—for testing companies, sponsor organizations, universities, or as consultants or service providers to the testing industry. The chapters provide examples and case studies of how various process tools have been applied to the testing industry. Readers from the measurement community should more readily understand the applications of quality to testing from these examples than from manufacturing and service examples.

This book is intended for both academics and practitioners in the fields of education, certification, personnel testing, and process improvement. Those interested in how the public and sponsoring agencies can encourage vendors to improve quality may also be interested in *Improved Testing*. The book also presents an international perspective, especially in the section on standards.

Improved Testing consists of 18 chapters, divided into six parts described briefly in the rest of this preface. A more detailed discussion of the contents of these chapters is provided in chapter 2.

Part I: The Quality Model

Part I provides an overall orientation to quality management. In chapter 1, Cheryl Wild provides an overview to the issue of quality in the testing industry. She describes the evidence for concern about quality in the testing industry, defines the term quality, describes the testing industry in process terms, and describes the cost of poor quality in the testing industry. In chapter 2, Rohit Ramaswamy presents the conceptual framework around which the chapters in the book are organized. Overall, a company that is effectively managing for quality uses three types of techniques: standards, design and planning, and continuous improvement. To be most effective, tools and techniques are linked by an engaged and enabling leadership. The following sections address leadership and the three types of techniques.

Part II: Leadership

Effective implementation of an enterprise-wide quality initiative requires the active involvement and engagement of leadership. In chapter 3, Richard Smith discusses the critical role of leadership in developing a culture that can successfully implement a continuous improvement initiative. The discussion includes leadership behaviors that reinforce and support the importance of quality, honest and open communication, rewards and recognition consistent with the improvement initiative, developing and using process metrics, and providing coaching and training to support the quality initiative.

Part III: Standards

Chapters 4 through 8 deal with how standards may be used to improve the quality of testing. Standards are a way of communicating how work will be done and what the expected outcome will be. An organization usually develops internal standards and metrics for managing processes and often adopts standards that have been developed by an external group as well. In chapter 4, Cheryl Wild and Joan Knapp discuss 25 external standards used by testing organizations. Their chapter describes three kinds of standards: those that apply specifically to testing organizations and have no third-party review component, those that apply specifically to testing programs with a third-party review component, and generic business standards used in the testing industry with a third-party review component. In chapter 5, Roger Brauer, using the Board of Certified Safety Professionals (BCSP) as an example, describes the value that accreditation provides to the various stakeholders of the certification process. Peter Kronvall, in chapter 6, provides the perspective of a Swedish organization that grants accreditation. Europe began applying the precursor of ISO/IEC 17024, a standard for bodies that certify people (as opposed to certifying products), in the mid-1990s. Using case studies of three certification programs he describes lessons learned in how certification bodies and the accreditation body can work together to improve the quality of the certification process. In chapter 7, Sharon Goldsmith and Michael Rosenfeld describe accreditation in the certification and licensing sector of the testing industry and discuss whether test vendor accreditation would be a worthwhile extension of the current accreditation procedures. In the final chapter of the standards section of the book (chapter 8), Marten Roorda provides an international perspective on existing quality standards for testing.

Part IV: Design and Planning

Chapters 9 through 12 address tools and techniques that help prevent errors before they happen. Standards tell a business what the accepted performance and practices are for the organization and they provide a baseline for expected quality. However, standards themselves do not ensure effective implementation. In chapter 9, Rohit Ramaswamy describes tools that can be used in designing a process to prevent or minimize the possibility of errors. Charles DePascale (chapter 10) presents best practices for a successful transition between vendors, a common problem in testing, since new vendors will have their own set of statistical programs, process steps, and internal standards that may differ from those of the old vendor. In chapter 11, Judson Turner provides the perspective of Georgia's Department of Education on requirements within a state department of education to ensure a successful assessment program, a quality testing division, a strong request for a proposal, and an effective contract negotiation. In chapter 12, Noel Albertson talks about some critical practical considerations in planning and organizing for a transition from paper-and-pencil testing to computer-based testing based on practical experience from the American Institute of Certified Public Accountants.

Part V: Monitoring and Improvement

Standards create a baseline for product and process quality. Planning helps to implement the standards effectively and to systematically design processes that will prevent errors and meet customer needs. However, no process works perfectly over time. The chapters in this section deal with tools and techniques for monitoring quality and improving it over time. Such activities are called process management or continual improvement. In chapter 13, Rohit Ramaswamy details seven steps for implementing process management in testing organizations.

David Anderson (chapter 14) describes the application of Six Sigma techniques at Educational Testing Service. He provides examples of specific projects and their results. Chapters 15 and 16 both deal with metrics or tools that can be used to monitor the testing process for security problems. David Foster, Dennis Maynes, and Bob Hunt describe the extent of cheating problems on standardized tests in chapter 15. They discuss data forensics or investigative methodology using statistical analysis of test scores and item responses for indications of cheating and piracy. In chapter 16, Ning Han and Ronald Hambleton describe a study comparing various item statistics that can be used to detect exposed test items in computer-based testing.

In chapter 17, Rick Dobbs and Stuart Kahl describe the development of integrated systems that support all the major processes associated with test development, test administration, scoring, and reporting processes. They suggest that the complexity of customer demands and the flexibility and speed that will be needed in the future cannot be met without the definition of key process metrics, and the use of integrated enterprise systems to track progress on these metrics and to allow effective management of the work flow.

Part VI: The Future of Quality in the Testing Industry

In chapter 18, Cheryl Wild and Rohit Ramaswamy tie together the results from the earlier chapters along two dimensions—first in terms of the quality triangle and, second, in terms of the processes in the testing industry. The earlier chapters in each part of the book cover distinct tools or techniques. The authors describe how organizations introduce an enterprise-wide system of process-driven quality using the entire triangle. In addition, the authors review the approaches currently in use in the testing industry, as described in the chapters and in their industry experience, and next steps needed by the industry.

Acknowledgments

The editors are grateful to Debra Riegert of Lawrence Erlbaum Associates, who supported this process from conception through the many steps to final product. Rebecca Larsen answered all our detailed questions with promptness and patience. The production staff improved the book immensely. Erin Burns designed the book cover with an artistic rendition of the quality triangle. We also want to thank our reviewer Stephen G. Sireci, University of Massachusetts.

We are also grateful to the chapter authors for contributing their time and expertise—there is never enough time to do everything, and we are especially grateful that *Improving Testing* became one of their priorities. We appreciate the willingness of our contributors to step beyond the concerns of proprietary self-interest and provide examples and case studies with the goal of improving the industry as a whole.

We would especially like to thank Ronald Hambleton for his mentorship in moving this from a concept to a book proposal. Nancy Petersen and Timothy Habick reviewed numerous drafts and were honest and free with their suggestions for improvement.

Finally, we would like to thank our spouses, Ed Bing and Shelia Pottebaum, for their patience and forbearance through multiple reiterations of "it's almost done."

Cheryl L. Wild, PhD
Rohit Ramaswamy, PhD

Part I

The Quality Model

1

The Risks and Costs of Poor Quality in Testing

Cheryl L. Wild
Wild & Associates, Inc.

Introduction

The premise of this book is simple—the testing industry as a whole is in dire need of improving the quality of its products and services. How can such a transformation occur? By applying process-oriented quality management tools and techniques along with psychometric principles to formally manage product and service quality. Not surprisingly, there are best practice organizations in the testing industry that are using some of these tools and techniques. In the chapters that follow, this book demonstrates how tried and proven tools used in achieving process quality in other fields have been applied by best practice organizations within the testing industry.

The purpose of this chapter is to provide an overall orientation to quality and process terminology and methods as they apply to the testing industry. First, the chapter discusses the evidence for whether there is a quality issue in the testing industry. Next, the term *quality* is defined and psychometric quality is described as a part of a more general definition of quality. Third, the testing industry is described in process terms, and finally, the costs of poor quality in testing are discussed. Is there really a quality issue in the testing industry?

Poor Quality or Poor Press?

Recent headlines have been lambasting test providers—"Report Blasts Tests for EMTs" (Scanlon, 2004), "Scores on Math Regents Exam to Be Raised for Thousands" (Arenson, 2003b), "Health Fields

Fight Cheating on Tests" (Smydo, 2003), "Errors on Tests in Nevada and Georgia Cost Publisher" (Hendrie & Hurst, 2002), "Scoring Error Clouds Hiring of Teachers" (Jacobson, 2004). From one year to the next the headlines are similar—the publisher, the sponsoring agency, and the test may change but the underlying issue remains the same: mistakes may have disastrous consequences for test takers.

For example, in the spring of 2000, a mistake in the key of the Minnesota Basic Standards Test resulted in 7,935 students being told they failed when they actually passed; 50 students did not receive their high school diplomas with their classmates (Rhoades & Madaus, 2003). In 2003, the New York Regents Math exam had such a high failure rate that the New York Commissioner of Education set aside the scores of juniors and seniors and established a panel to review the impact (Arenson, 2003a).

Certification and licensure tests have not been immune to problems. In 2004, approximately 4,000 examinees who had previously failed their teacher certification examination were informed that a mistake had been made and that they had passed rather than failed. Many of these examinees had lost opportunities to pursue teaching careers as a result of their initial failing scores (Jacobson, 2004).

Tests used for college admissions have also had their share of problems. A scanning problem resulted in incorrect scanning of some answer sheets for the October 2005 administration of the SAT, resulting in more than 4,400 incorrect scores (Caperton, 2006). Approximately 1,500 Advanced Placement Examinations were lost after the May 2006 test administration (Strauss, 2006).

Is this problem any more acute now than it was a few years ago? Absolutely. This is because the *number of tests* administered has increased dramatically, more tests have *high-stake* consequences, and the *changing platform* of test delivery (from paper-and-pencil tests to computer-based tests) has introduced new problems along with new delivery opportunities.

The No Child Left Behind (NCLB) law has resulted in a tremendous expansion of testing. NCLB will at least double the number of tests administered in elementary and secondary schools. This law requires that all states test math and language arts in grades 3 to 8 and in at least one year in high school. The law doesn't just impact student testing; states are required to have *qualified* teachers in the classroom. As evidence that teachers are qualified, states are also increasing the demand for teacher certification and/or licensing.

The number of tests administered in the business sector appears to be increasing as well, although there is no single source that collects this information. In certification testing, the growth appears to have an international impetus. The International Organization for Standardization, a worldwide organization of standards bodies, developed a new standard for organizations conducting the certification of persons (ISO/IEC 17024, 2003), and envisioned that the use of this standard would result in an explosion of certification and licensure tests in the global market place. By developing this standard, the organization hopes to establish the environment for mutual recognition and global exchange of personnel. Recognition as an international standard requires that a standard receive approval by at least 75% of the member bodies—reflecting the international importance of certification in the marketplace today. The standard provides general requirements (including organizational structure, finance, security, psychometrics, management system, subcontracting, human resource, and certification process requirements) for bodies operating certification of persons, allowing for a third party to verify that certification bodies are operating in a manner consistent with the standard. (More details about this and other standards are provided in chapters 4 through 8 of this book.)

Increased testing has placed a huge demand on the industry to increase production capacity. In manufacturing, increasing capacity requires building a factory or a plant. The testing industry is a knowledge-based industry—one that is very dependent on subject matter experts (to write questions, edit questions, and develop tests) and on psychometricians, usually PhDs with expertise in measurement theory and statistics. As Toch (2006) discovered in his review of the education testing industry, the testing industry is having a great deal of difficulty finding qualified people to meet the increasing demand.

Not only have the numbers of test takers in both the education and certification/licensure arenas increased—more of the tests have *high-stake* consequences, especially for schools, students, and teachers. Test results trigger consequences for schools—if students don't achieve annual yearly progress (AYP) goals, schools may be required to report the failure to parents, develop improvement plans, allow students to transfer to other schools or districts, or provide resources for tutoring. After four consecutive years of not meeting goals, the state could withdraw funding, require personnel changes, and/or require curriculum revisions. Students may not be allowed to graduate in some states if they don't pass the high school graduation test. And teachers' jobs may depend on passing a certification examination or on improving their students' scores. In some states, including Alaska, Florida, and Texas, bonus pay for teachers is being linked to improvements in student's test scores (McNeil, 2006).

The increasing importance of test results has implications for testing products and services. Specifically, the higher the stakes for any given test the greater the motivation for cheating. More attention may need to be paid to the security of test storage, to training test administrators in their responsibilities for maintaining security, to checking data for evidence of cheating, and to providing new test editions more often. Processes and procedures that seemed adequate even a few years ago may no longer meet current user needs.

Finally, not only are more tests being given with higher stakes, but there is also a *changing platform* for test delivery. The testing industry is in transition from *paper-and-pencil testing to computer-based testing* in both the educational and business settings. This transition requires the development of more complex systems and software, new procedures for registering for the examination, new methods for score reporting and equating, new methods for selecting questions for the test, and new testing sites. These changes are creating new problems while eliminating others. For example, one of the problems with reporting educational test scores in a timely manner is the time required to have paper-and-pencil tests returned from a test center. Computer-based testing eliminates the security and logistical issues connected to shipping test booklets and answer sheets. However, new problems can and do occur—a loss of power during a testing session or loss of an Internet connection would seriously disrupt the test session and might even mean a loss of examinee responses. The development of a new platform requires design of new processes and procedures to ensure quality and reliability of delivery.

These three changes in the environment—drastically increased volume, changing customer requirements in terms of quality, and a changing delivery platform—would strain and challenge any industry. The evidence of problems in the delivery of products and services in testing strongly suggests that the testing industry is indeed challenged and that there is a need for a quantum leap in quality for the testing industry in the next few years.

There is little evidence, however, that quality is on the rise. For instance, Kathleen Rhoades and George Madaus (2003), researchers at the National Board on Educational Testing and Public Policy at Boston College, identified over 100 errors that occurred in testing programs over the last 25 years. Each error was described in detail in appendices to the report, allowing for independent evaluation of the errors. The causes of the errors (primarily ambiguous questions, incorrect scoring, score equating errors [equating is an empirical procedure to produce scores on multiple forms of a test that can be used interchangeably], programming errors, and data management errors) have not changed across the 25 years of the study. And these are only a subset of the errors that can occur—problems in administration at the test center, delays in score reporting, or breaches of security are not reported in the study.

Toch (2006) has reviewed the current status of the educational testing industry and has identified problems in industry capacity, timeliness of delivery, accuracy, and uneducated consumers (consumers include state department of education staff who contract with testing companies).

At the same time as newspaper headlines decry the problems with testing, testing experts claim that there are no serious quality problems in the testing industry. For example, Cizek (2001) argues that tests are better than ever:

> High stakes tests have evolved to a point where they are: (a) highly reliable, (b) free from bias, (c) relevant and age appropriate, (d) higher order, (e) tightly related to important, public goals, (f) time and cost efficient, and (g) yielding remarkably consistent decisions. (p. 25)

As reported in chapter 6, Goldsmith and Rosenfeld (2007) interviewed 15 individuals, 10 representing the testing community. They found that "interviewees across all industries commented on the fact that there were no blatant problems in the testing community" (p. 137). They go on to say "They [existing standards] appear to be working well and although there have been isolated instances of errors in scoring and breaches in security, there have not been any reported wide scale abuses within the industry" (p. 137).

Why this schism in opinions about the quality of testing? This is because the users of tests and the developers of tests are talking about different aspects of quality. Developers of tests often talk about quality of testing in terms of psychometric considerations, like construct validity, score reliability, and content bias. Users of tests look at quality from a broader perspective. Users usually assume (possibly erroneously) that the psychometric quality is built into the test. Test users are more likely to ask: Did the test books arrive on time? Were all the materials ordered included in the shipment? Were the boxes delivered to the right location, on time, and undamaged? Were there typos or more serious errors in the test book or the test administration handbook? Were any answer sheets lost in transit? Were the score reports accurate and delivered in a timely manner? These are important aspects of quality to test users, but not necessarily a primary consideration when discussing the psychometric quality of a test.

If the psychometricians and testing experts don't recognize a quality issue in the testing industry, is the press exaggerating the issue? Can both be correct? Yes. It is likely that the disparity of opinion about quality lies in the lack of specificity in what is meant by the term quality.

What Is Quality?

In our definition, testing quality must refer to two components of the testing experience—the quality of the testing products (the test itself, test books, test administrator manuals, scores, and reports) and the quality of the services related to the testing products (ordering of test books, packaging, administration of computer-based or paper-and-pencil tests, return of test books and answer sheets, scheduling of a computer-based test, customer service, and score verification).

Juran (1999) divides quality into two aspects: features that meet or exceed customer needs and freedom from deficiencies. These two aspects of quality apply to both testing products and testing services.

Table 1.1 presents some examples of product features that meet customer needs and examples of correct outcomes that evidence a lack of deficiencies. One of these examples is for the test score (a testing product). A desired feature of a test score might be that total scores are comparable from one edition of the test to another. Reporting raw scores on different editions of a test is relatively easy—you just add up the number of correct answers. Users, however, have difficulty in making judgments from raw scores when multiple editions of a test are in use—you cannot tell whether a 75 on edition A means the same thing as a 75 on edition B. Adding the feature of comparable scores—when a 75 means the same thing on editions A and B—requires equating.

TABLE 1.1   Examples of Two Aspects of Quality for Products of the Testing Industry

	Examples of Two Aspects of Quality	
Product	Features That Meet/Exceed Customer Needs	Freedom from Deficiencies
Test Book	• Color pictures • Test books color coded for ease of handling • Appropriate size of font in test book • Items in the test are new, secure • Test specifications and content designed to meet psyschometric standards (e.g., for certification test, test specifications based on job analysis)	• No typos • Pages correctly collated • Directions in correct place • On-time delivery to testing center
Test Score	• Scores of multiple test editions reported on same scale • Subscores provide additional information for test takers to improve performance	• Correct keys used for scoring • Responses accurately scanned • Equating meets standards of testing industry, psychometrically correct • Test scores accurately matched to examinee files
Score Report	• Customized for different users—e.g., parents, teachers, students, test takers, association management, industry reports, government reports • Available electronically as well as on paper • Available in multiple languages for parents who may not speak English	• Accurate scores/summary data • Reports prepared on time • Reports delivered to the correct person/location

Equating requires a design for data collection and a method for transforming scores from two or more test forms to a common scale. For example, equating often relies on an administration procedure that uses common items in two forms or random/spiraled administration of the two forms to test takers. The results are then statistically analyzed, and scores from the two tests are reported on the same scale. Deficiencies in a test score might result if the scoring is done incorrectly (possible causes would be using an incorrect key or if the scanning of answer sheets isn't picking up the answers correctly), if test scores are mismatched to examinee files, or if the equating is done incorrectly. Lack of deficiencies would mean that scores are calculated correctly, that equating is applied correctly, and that scores are correctly matched to examinee files.

Table 1.2 presents examples of test services that meet customer needs and examples of correct outcomes that evidence a lack of deficiencies. The delivery of test books is a service provided by testing vendors. Possible deficiencies in the delivery of test books

TABLE 1.2 Examples of Two Aspects of Quality for Services of the Testing Industry

Services	Examples of Two Aspects of Quality	
	Features That Meet/Exceed Customer Needs	Freedom from Deficiencies
Ordering of Testing Materials	• Customer service department available to answer questions about ordering • Quick turnaround after material is ordered • Online ordering easy to access and use	• Directions clear and correct • Ordering available well before test administration
Delivery of Test Materials	• In school district, test materials are packaged by school • Delivery to specified person only, to assure security	• Test materials ordered are materials delivered • On-time delivery • Test books delivered to correct location
Training of Test Administrators	• Online training of test administrators with "certification" test • Test administration manual provided in advance for administrators to study	• Test administration manual directions match directions in test book • Training available before testing begins • Test administrators understand and implement secure procedures

are late or missing deliveries, deliveries going to an incorrect location, or deliveries that do not match the test order.

In the school setting, test booklets are typically sent to a central point in the school district rather than directly to each school. An extra service is to package booklets and related material by school within district. This service is typically more expensive than packaging all the district's materials together and requiring school personnel to sort the materials. Providing such a service often increases customer satisfaction, even though it is at an increased cost to the school district.

This service feature is a good illustration of how a characteristic that was once a feature that may meet or exceed a customer need can become an expectation—the lack of which can be viewed as a deficiency in the eyes of the customer. In New Jersey, the statewide assessment tests are delivered to districts in packets by school. Schools expect this. In the first year of testing with a new assessment for English language learners, the agreement with the vendor was to send test booklets to the district but only to package by school for the very large districts. This method of packaging seemed reasonable because it kept costs down and because the numbers of students and test books were much smaller than in the statewide assessment program. Districts, however, complained about the poor service, even though the contractor was delivering the service as promised. In the eyes of the customer, this was a deficiency.

In this book, the notion of testing quality covers three key ideas. The first is that quality cannot be defined without the customer of the testing process in mind—ultimately, they are the ones who experience the test, and whose satisfaction determines our success. The second is that all work in a testing organization is done through processes, and that an end-to-end view of these processes is needed to understand where the organization needs to focus to satisfy its customers. The third is that whereas quality improvements may involve costs, it is also costly to have poor quality, and the costs of process improvements must always be balanced against the consequences of not doing anything at all. These three ideas are discussed sequentially in the following sections.

Focus on the Customer

Quality is in the eye of the beholder—it is important to view quality from the eyes of the customers of the testing process and not just from the viewpoint of those who design and create the tests. A high-quality test is one that results in high customer satisfaction. Deficiencies increase customer dissatisfaction (customers expect defect-free products) whereas features can increase customer satisfaction, even when customers must pay more for the feature. Overall customer satisfaction requires that both products and services have the desired features and be delivered free from deficiencies to the end customer. To achieve this, though, requires recognition of the needs of all the customers involved in the development and delivery of the product.

Customer satisfaction requires that the multiple customers of the testing industry be recognized and included in the design and delivery of tests and services. Test takers are obviously customers; parents are sometimes customers. Test sponsors often hire a testing company, so the test sponsor is a customer. Test sponsors, however, may hire several vendors in the delivery of a complex testing program. For example, the College Board currently hires the Educational Testing Service (ETS) to develop questions for the SAT and Pearson Educational Measurement to score the SAT. Pearson Educational Measurement is thus the customer of ETS—Pearson uses the scoring keys and rubrics provided by ETS. Producing a quality product for the test taker requires that the multiple vendors and departments within each organization work well together. Total customer satisfaction involves satisfying all these customers: this requires considering quality in all steps of the process of designing and delivering testing products and services. This is a much broader definition of quality than is normally considered by testing experts.

In summary, quality in the testing industry requires the delivery of test products and related services that are defect-free and provide the product and service features that are desired by customers. This requires the design of effective processes and procedures that will achieve the desired level of quality. Understanding the process perspective is a critical requirement for doing this.

Understanding Processes

All products and services in an organization are created through processes. A *process* is simply a series of steps that take an input and produce an output. Figure 1.1 illustrates an example of a business process in the testing industry—the test creation process. The inputs

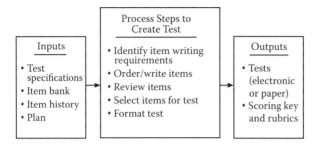

Figure 1.1 Test creation process.

for test creation include the specifications for the test, the item bank and item history, and planning information such as due date, budget, and so forth. A detailed series of steps (the process) are executed by the test developers to take the existing items and test plan and create a new test (the output). The output will be a test (paper-and-pencil or computer file for computer-based delivery) along with the scoring key and any scoring rubrics.

Companies with a process approach to managing their business identify important business processes that affect customer satisfaction, define the steps involved in the processes, specify how the processes will fit together, and manage the performance of the processes. Organizations that view all their work through this process lens are better able manage the complex interrelationships between the different work activities and to identify where these activities touch the customers. This helps them to focus on areas where quality is most critical. Organizations that view their work as functional *silos* are less able to identify where they have an impact on their customers, and consequently are less able to determine how the quality of their products and services can be improved.

Generically, business processes typically fall into two categories: *core* and *enabling* processes. Core processes are those that are typically critical to customer satisfaction. Organizations typically devote a higher level of attention to these core processes. Enabling processes are also critical to the operation of a business but may not pertain to delivering the product or service to customers. Another way to look at the difference between core business processes and enabling processes is that core business processes are important to the external customer needs, and support processes are important to internal customer needs.

TABLE 1.3   Major Business Processes of the Testing Industry

Core Business Processes
1. New Product Development
2. Test Creation
3. Test Ordering, Delivery, and Administration
4. Scoring and Reporting
5. Customer Relationship Management and Communication

Key Support Processes
6. System Development and Support
7. Quality Management
8. Security Management
9. Human Resources
10. Billing
11. Marketing
12. Finance

How can the idea of business processes be translated to the testing industry? Table 1.3 presents a list of common processes in testing organizations. New product development, test creation, test ordering/delivery/administration, scoring and reporting, and customer relationship management and communication are examples of core business processes in a testing organization. Support processes might include system development and support, quality management, finance, security management, and human resources.

In reviewing the business processes in the testing industry there are two important facts to recognize—the core process will differ for different organizations, and a support process in one organization may be a core process in another. Systems development and support is one process whose importance varies considerably from one organization to another. For example, it might be a support process for testing organizations that primarily develop tests delivered by an outside computer-based test delivery vendor. In such an organization, the systems processes of item banking, scoring, statistical analysis, billing, and so forth, may be primarily used internally by the organization and not visible to the external customer. In the educational testing arena, products and services are becoming more and more customized. Score reports must be customized to the state and sometimes the district level. In an organization that provides extensive customized reporting or delivery of computer-based tests, system development and support may be a core process.

Of the other support processes, security management and quality management (sometimes including internal auditing) may not have been identified as important processes historically, but are becoming more important. Examples of other support processes include human resources, marketing, finance and billing, and public relations. Whether these processes are important or not in an individual testing organization will depend on the organization.

Why is it important to understand that a testing organization is made up of a number of specific processes and to identifying the core business processes? Identifying the core processes and relating them to the organization's strategic plan helps an organization identify where to invest resources to improve the quality of its products and services. Once an organization identifies important business processes, the next step is defining the process steps in detail. Defining business process steps not only allows an organization to have a guide for its employees about how to do the work, it also improves the likelihood that work will be conducted in a consistent manner and at a consistent level of quality. It enables an organization to identify hand-off points where quality may be compromised, allowing the organization to systematically apply quality tools to improve quality. In chapter 13, Dr. Rohit Ramaswamy describes in detail how an organization would go about documenting and managing their processes.

At this point in the discussion, the reader is probably beginning to worry about costs. The testing industry is not one with abundant resources. Taking a test is like having a root canal or taking vitamins—you do it because you have to or because someone told you it would make you healthier—you don't want to spend any more on a test than necessary. Test sponsors are often not-for-profit organizations or government organizations that are not necessarily funded to pay for additional expenses. How much will a quality focus cost the industry? In order to answer this question, it is important to understand the implications of not having a quality focus. A succinct way of doing that is through understanding the cost of poor quality.

Cost of Poor Quality

There is no doubt that redesigning work processes and management processes to improve quality has cost implications. The real question is whether or not there will be a return on the investment. Even not-for-profit organizations must have a positive cash flow.

The costs of poor quality include costs of additional quality control activities, rework, customer compensation, or litigation that may be needed when a test product or service of poor quality is produced. These costs are currently built into the cost of producing a test. How much of the expense of a test is related to poor quality? Gryna (1999, p. 8.10) indicates that even using conservative methods of calculating cost of poor quality, "in most companies, the cost of poor quality is a large sum, frequently larger than the company's profits." Gryna reports that cost of poor quality estimates range from 10% to 30% of sales, or 25% to 40% of operating expenses. Although these estimates are not testing-industry specific, it is likely that the testing industry falls within this range.

Cost of poor quality is computed by totaling the costs in four categories:

1. Prevention (e.g., quality planning, process planning, employee training for work, process capability studies)
2. Appraisal (e.g., quality control, quality audits, cost of test equipment, training for inspection/quality control, and quality reporting)
3. Internal failures (e.g., cost or rework, design failures, scrap)
4. External failure (e.g., customer complaint resolution, costs associated with legal suits, public relations)

Although the total cost is estimated in different ways in different studies, the estimate of total quality cost is often felt to be conservative because it cannot measure the lifetime costs of lost sales due to poor quality or bad reputation.

Once the total cost of poor quality is estimated, the percent of the total due to each of the four categories is obtained. Typical cost of poor quality studies show woefully low amounts of quality costs dedicated to prevention (0.5% to 5.0%), higher amounts allocated to appraisal, and failure costs several times appraisal costs. In a case study applying the cost of poor quality metrics to several subprocesses in the testing industry, Wild and Kovacs (1994) found that the proportion of quality costs to prevention, appraisal, and failure was very similar to the typical figures in other industries.

Most companies do not make public the cost of poor quality, and there are no reports estimating the overall cost of poor quality in the testing industry. There is, however, every reason to suspect that the cost of poor quality is high. Typos, multiple-choice questions without a single best answer, vague questions, and missing pages are examples of deficiencies in a test book. Test publishers have extensive quality control procedures in place to prevent deficiencies such as these from appearing in published test books (or for that matter, in computer-based tests). Many of these quality control activities involve "checking the checker," that is, revalidating the results of previous inspections because, in many cases, the organization does not have test creation processes that are designed to minimize or prevent errors. The logic that companies use is that although the quality control to prevent deficiencies is costly, it is worth the expense because the costs and risks incurred when deficiencies escape the in-house quality control procedures and appear in a student's test are very costly for the test publisher. A better option, however, might be to invest in the creation of processes that reduce errors by providing systematic procedures and tools to aid test creation. This will result in increased confidence not only among the test developers, but among test consumers as well.

If readers believe that the testing industry is a monopoly and that customer confidence isn't relevant, they need only look at the College Board. The College Board is experiencing a slump in customer confidence as a result of the October 2005 scoring mishap mentioned earlier in this chapter. Klein (2006) interviewed various admissions officials about the drop in scores on the revised SAT, and quoted one admission official as saying, "There's been much less of the trust of the College Board and the reliability of the SAT" (p. 18) as a result of the October scoring problem. The number of test takers for the SAT is down slightly at the end of the testing year, whereas the number of test takers for its competitor, ACT, is up (Honawar & Klein, 2006). More schools are willing to accept either the SAT or the ACT than in the past, leaving the choice of test to the test taker. Although the trend for growing volumes on the ACT existed prior to the College Board scoring problem, it is fair to speculate that the negative press of the SAT scoring incident encouraged some test takers to move to the ACT and encouraged some admissions officers to consider ACT as a viable alternative to the SAT.

Adding to the cost of poor quality is the cost of fines when vendors do not meet their commitments, the costs for insurance against errors, and the costs for settling lawsuits when errors reach court. For example, an $11.1 million fund was set aside to pay damages to the thousands of teachers who were given incorrect licensing exam scores in 2003 and 2004 (Pope, 2006). The contract with the vendor for the Illinois State Achievement Examination (ISAT) included damages, and the exact costs to the vendor for problems in delivering the 2005–2006 school year had not been determined when Colindres (2006) reported that these damages may be in the ballpark of $1.6 million.

What is the impact of the cost of poor quality on the cost of testing for an individual test taker? Suppose a typical student applying for graduate school took the GRE General Test ($130) and the GRE Mathematics test ($150). The student requested that the Mathematics test be hand scored ($30) to make sure the answer sheet was scanned properly, and then took advantage of the question and answer review service ($50) for the General Test. Finally, the student requested three additional score reports be sent ($45). Overall the bill for testing would be $405, and assuming cost of quality of 10 to 30 percent of the sale, from $40 to $120 of that bill would be related to the costs of poor quality.

The reason for quantifying these costs is to determine whether the costs involved in improving work processes are justifiable. Can the testing industry afford to implement major quality initiatives? Given that the cost for poor quality is so high, the author's opinion is that the cost is justified. So how do organizations move beyond the inspection and improve processes? This book is organized around three principles that, when used together, can result in transformational changes in organizations. These three principles—improving quality through implementing standards, preventing errors through planning and design, and implementing continual improvement—and how the book is organized around the three principles are discussed in detail in chapter 2.

It is important to clarify here that the principles and practices of quality described in this book are applicable to all organizations, not just large ones. Examples and experiences mentioned in this book have come from larger testing organizations like Educational Testing Service to smaller associations like the American Institute of Certified Public Accountants and the Board of Certified Safety Professionals. The tools can be used by all organizations, big or small, and in the spirit of continual improvement, by each individual in an organization to improve the quality of work in his or her process step. The ingredients for success: good leadership, conformance to standards, documented and consistent processes, the use of metrics to monitor quality, and focus on customer satisfaction are the same irrespective of organizational size. The only difference between big and small organizations may be the way in which a quality improvement program gets deployed; smaller organizations might seek to implement projects that can be accomplished quickly and with fewer resources. We describe different deployment options in the last chapter of this book.

Conclusions and Lessons Learned

The testing industry as a whole is having problems meeting customer requirements. *Improving Testing* provides examples and discussion of how the use of process-oriented quality management tools and techniques along with psychometric principles can be used to transform quality in the testing industry. To provide a context for the rest of the book, this chapter defines quality, describes a process, discusses the core processes involved in the testing industry, and discusses briefly how cost and quality are related.

In summary:

- Improving testing requires improving both the products and services provided by the testing industry.
- Providing quality testing includes the addition of features that meet or exceed the customer needs without deficiencies.
- Adding features to products and services generally costs money but may increase business by improving customer satisfaction.
- Eliminating deficiencies generally decreases cost. Deficiencies that become visible to the customer generally decrease customer satisfaction. Customers expect companies to deliver products without deficiencies, so a lack of deficiencies does not usually improve customer satisfaction.
- Cost of poor quality in the testing industry may account for 10% to 30% of testing sales, or 25% to 40% of operating expenses. With these magnitudes of possible savings, the industry can surely afford to invest in improving quality.
- Defining testing processes, managing processes, and designing and improving these processes can result in major improvements in quality.

The testing industry is at a crossroads. Public awareness of testing and quality issues is at an all-time high. Individual companies must choose between continuing to address quality concerns in the same old way and applying proven process tools to improve testing. The risks of not addressing quality concerns to testing companies are great:

- Individual companies that cannot meet customer quality requirements will cease to exist or be acquired by other organizations.
- Cost and quality concerns may result in government or consumer groups dictating quality requirements or creating oversight bodies.
- Poor quality of educational testing could inhibit the ability of schools to improve education based on testing metrics.
- Inability of U.S. organizations to meet international accreditation standards could reduce the acceptance of U.S. credentials internationally.

Chapter 2 provides a framework for organizing the various tools and techniques that will help improve the testing industry.

References

Arenson, K. W. (2003a, August 27). Early trials of Regents test foresaw failure rates at a high rate. *New York Times*. Retrieved October 7, 2004, from http://query.nytimes.com

Arenson, K. W. (2003b, August 30). Scores on math regents exam to be raised for thousands. *New York Times*. Retrieved October 7, 2004, from http://query.nytimes.com

Caperton, G. (2006, April 6). We worked day and night. *USA Today*. Retrieved April 4, 2006, from www.usatoday.com

Colindres, A. (2006, May 19). Harcourt agrees to pay $1.6 million for ISAT foul-up. *The State Journal-Register Online*. Retrieved May 25, 2006, from http://www.sj-r.com

Cizek, G. J. (2001). More unintended consequences of high-stakes testing. *Educational Measurement: Issues and Practice, 20*(4), 19–27.

Gryna, F. M. (1999). Quality and costs. In J. M. Juran & A. B. Godfrey (Eds.), *Juran's quality handbook* (3rd ed.). (pp. 8.1–8.26). New York: McGraw-Hill.

Goldsmith, S. M., & Rosenfeld, M. (2007). Accreditation in the certification and licensing sector of the testing industry: Current status and future possibilities. In C. L. Wild & R. Ramaswamy (Eds.), *Improving testing: Applying quality tools and techniques to assure quality* (pp. 125–144). Mahwah, NJ: Lawrence Erlbaum Associates.

Hendrie, C., & Hurst, M. (2002, September 4). Errors on tests in Nevada and Georgia cost publisher Harcourt. *Education Week*, p. 24.

Honawar, V., & Klein, A. (2006, August 30). ACT scores improve; more on east coast taking the SAT's rival. *Education Week*, p. 16.

International Organization for Standardization. (2003). *International standard ISO/IEC 17024 conformity assessment—general requirements for bodies operating certification of persons.* Geneva, Switzerland: Author.

Jacobson, L. (2004, July 28). Scoring error clouds hiring of teachers, *Education Week*, pp. 1, 14.

Juran, J. M. (1999). How to think about quality. In J. M. Juran & A. B. Godfrey (Eds.), *Juran's quality handbook* (5th ed.). (pp. 2.1–2.18). New York: McGraw-Hill.

Klein, A. (2006, September 6). Score drop prompts debate over effects of revised SAT; girls outdo boys in writing. *Education Week*, p. 18.

McNeil, M. (2006, September 6). States giving performance pay by doling out bonuses. *Education Week*, pp. 30, 33.

Pope, J. (2006, March 14). Test service reaches deal in error suits. *Seattle Post-Intelligencer.* Retrieved March 16, 2006, from http://seatlepi.nwsource.com

Rhoades, K., & Madaus, G. (2003). *Errors in standardized tests: a systemic problem.* [Monograph]. Boston College: National Board on Educational Testing and Public Policy. Retrieved from http://www.bc.edu/research/nbetpp/statements/M1N4.pdf

Scanlon, B. (2004, May 8). Report blasts tests for EMTs. *Rocky Mountain News (CO).* Retrieved October 7, 2004, from http://nl.newsbank.com

Smydo, J. (2003, August 3). Health fields fight cheating on tests. *Post-Gazette.* Retrieved September 19, 2006, from www.post-gazette.com

Strauss, V. (2006, September 2). Hundreds more AP test scores reported missing. *Washingtonpost.com.* Retrieved September 7, 2006, from http://www.washingtonpost.com

Toch, T. (2006). *Margins of error: The education testing industry in the no child left behind era.* Retrieved February 11, 2006, from www.educationsector.org/usr_doc/Margins_of_Error.pdf

Wild, C. L., & Kovacs, C. (1994). Cost-of-quality case study. *Rediscovering quality: The next step* (pp. 3B-19–3B-29). Conference proceedings of the Juran Institute.

2

A Framework for Testing Industry Quality Improvement

Rohit Ramaswamy
Service Design Solutions, Inc.

Introduction

The purpose of this book is to describe tools and approaches to avoid errors in any aspect of testing that affects test takers and that results in loss of reputation or financial liability to the testing company and testing sponsors. In the chapters that surround this one, authors with diverse backgrounds and experience describe the current situation in the testing industry and suggest methods by which errors can be avoided or reduced. This chapter provides a framework for a holistic approach to quality improvement and describes how the chapters that follow are organized around the framework. This framework and the key ideas around each of its four components are discussed to help readers synthesize the ideas and viewpoints of the different authors who have contributed to this book.

Before we introduce the conceptual framework, it is useful to elaborate on what is meant by *errors*. When most of us think of errors, we tend to think in terms of *mistakes*. In day-to-day conversation, and in reports in the press, we hear about testing errors primarily as mistakes in test scoring or reporting that resulted in incorrect outcomes for test takers. Even within testing companies, the notion of error reduction is often confined to inspection and control activities that are needed to make sure that no mistakes are made in item review, test development, key assignment, and score reporting. In this book, however, we take a broader view of errors. The purpose of this book is to describe tools and techniques to improve testing *quality*.

As mentioned by Dr. Wild in chapter 1, the characteristics of high quality testing go beyond just mistake-free tests. Quality of tests refers both to the quality of the testing products (test items that meet high standards, test books that give clear directions, score reports that provide meaningful feedback, etc.), and also to the quality of the testing processes supporting the products (ordering of test books, packaging, test administration, delivery of computer-based tests, etc.).

Quality can refer to meeting the needs of the customer and avoiding problems. Later in *Improving Testing* DePasquale reports on the potential problems involved in transitioning from one test contractor to another. One potential problem is that contractors have their own analysis programs and procedures to do nominally the same work, yet these may produce different results. As a consequence, the quality may be compromised without any overt error being made. Security is another topic discussed later in the book. Poor security can result in some examinees having an unfair advantage over others and compromise the validity of the scores. In our definition, all these quality attributes (inconsistent analysis procedures or inadequate security) can be categorized as errors.

This definition of errors is akin to the definition of *defects* in common process improvement methodologies such as Six Sigma (see, for example, Pande, Neuman, and Cavanagh, [2000]), or in the Juran (1999) definition of quality as *freedom from deficiencies*. In these methodologies, a defect is defined as an occurrence when a customer or business requirement is not met, resulting in customer dissatisfaction. Depending on the requirements, defects can be defined as timeliness, accuracy, cost, or any other quality attribute of interest. The chapters in this book focus on multiple aspects of quality.

Conceptual Framework for Test Quality Improvement

The conceptual framework for test quality improvement that we employ in this book is the *test quality triangle* shown in Figure 2.1. The three legs of the triangle are *implementing standards* for quality, *planning and design*, and *monitoring and improvement*. At the center of the triangle is *leadership engagement*.

The quality triangle, with its enclosed circle, represents the core ideas that we are trying to promote in this book. Many testing organizations are working on one or more legs of the quality triangle, or leadership, but few if any organizations are working on all three legs and on the center in an integrated and consistent way. For example, some organizations may be working on achieving accreditation;

Figure 2.1 The quality triangle.

other organizations may be launching Six Sigma and similar process improvement methodologies, while still other organizations may be working on integrated process management or on leadership issues. But the key point that we want to emphasize in this book is that each of these activities, conducted alone, can only partially contribute to the overall quality of the testing organization. An organization that truly seeks to be defect free needs to use the quality tools and methods described in this book simultaneously, across the organization, and integrate their use with the strategic directions of the organization. The last chapter in this book describes how this can be achieved through the strategic deployment of a quality program in a testing organization.

The next sections of this chapter describe the quality triangle in greater detail and discuss how the chapters in this book contribute to the discussion of the key aspects of each leg of the triangle and of the leadership circle.

Beginning with Leadership

At the center of the quality triangle is *leadership engagement*. A successful quality improvement initiative in a testing organization is not simply a matter of putting some tools and methods together and running a few projects. An organization that is committed to error-free processes that consistently meet customer requirements not only needs effective technical implementation, but also needs to accept the discipline around designing and maintaining error-free processes as a part of the organization's culture.

Experience with quality deployments in various industries (Smith & Blakeslee, 2002) has shown that this often requires a fundamental transformation of the organization's culture, and the testing industry is no exception. The transformation needs to be championed and led by the organization's senior management team, and formal change management and communication practices need to be introduced to support the organization's quality improvement initiative. Leaders of quality improvement efforts face the dual task of taking tough steps to fight organizational resistance to change, while at the same time being coaches and guides as the company moves through uncertain terrain.

A survey conducted by PricewaterhouseCoopers (Smith & Blakeslee, 2002) determined that senior management behavior coupled with open and honest communications are the most important success factors for effective organizational transformation. The role of leadership in achieving successful and sustainable improvements in test quality therefore cannot be underestimated.

In chapter 3 of this book, Richard Smith describes the leadership elements that drive successful performance through a quality improvement implementation. Mr. Smith identifies the critical success factors for organizational change and leadership best practices that are most successful in guiding organizations through that change. Mr. Smith concludes that the key factors for achieving successful change in the testing industry, such as strong and committed leadership, good communications, appropriate reward and recognition systems, relevant metrics, and coaching and skill building, are not different from those in other industries. He suggests that as the testing organizations make a commitment to quality improvement, they must simultaneously embrace need for leadership understanding and expertise in leading change and ensure that it develops or hires engaged leaders who can ensure that these success factors are in place. Otherwise, the attempts to improve test quality will only produce sporadic and uninspiring results.

Implementing Standards

The first leg of the test quality triangle is *implementing standards* for quality. *The Standards for Educational and Psychological Testing* (1999) immediately come to mind in any discussion of standards in the testing industry. In the discussion of standards for improving quality, however, a much broader concept of standards is necessary. Standards are a way of communicating how work will be done and what the expected outcome will be. They are tools that helps assure that, when work is done by different people, the product will be nearly the same each time. They relate not just to the psychometric requirements for testing, but also to the supporting or sustaining processes described in chapter 1. The use of standards implies that there is a way of measuring or evaluating whether the standards are being met.

There are many dimensions across which standards relevant to the business of testing may vary. In chapters 4 through 8 of this book, standards are compared across a number of dimensions:

- Standards may be internal to an organization or standards designed through a consensus process, external to an organization.
- The content of the standard can be testing-industry specific or industry neutral and process specific (e.g., quality management system process or software development process).
- Some standards include a third-party review component, whereas others are more guidelines or suggestions.
- Some standards include the requirement of continuous improvement, whereas others specify minimum requirements without mentioning quality improvement.
- Some standards are recognized nationally, whereas others are developed and approved through international consensus.

Although all of the external standards identified by the various authors (26 or more) are important to the quality of testing, certain aspects of implementing standards are more important than others in an integrated approach to improving the quality of a testing organization.

First, although standards specific to the testing industry are necessary, they are not sufficient to ensure quality. For example, several testing organizations are using standards such as SysTrust™ and ISO/IEC 27011:2005 (see chapter 4) because they do not believe the testing-specific standards contain sufficient detail on system security and reliability to judge the effectiveness of their processes in these areas. Roorda (chapter 8) also points out that testing standards do not often cover the entire range of the testing processes.

Second, the use of standards and accreditation (or third-party review) is an entry into continuous quality improvement in a testing organization. In chapter 5, Brauer describes how, in the process of achieving and maintaining three different third-party accreditations, the Board of Certified Safety Professionals (BCSP) became a process-oriented organization, implemented process metrics, and continuously improves processes. External audits, even when the organization complies with the standard, sometimes suggest areas for improvement. In chapter 6, Kronvall describes how the use of accreditation standards for bodies who certify people in Sweden has resulted in improvements in both the process of providing accreditation and the processes used by the certification bodies.

Third, unless there are consequences to the use of standards, as Marten Roorda describes in chapter 8, they may not be consistently used and applied. In chapter 8, Roorda describes how the Dutch COTAN reviews tests and provides a public rating of them. As a consequence, CITO has identified a minimum quality metric on the COTAN rating for their tests. Is it likely that increasing consequences of meeting or not meeting standards can be applied to testing organizations in the United States?

In the United States, some small number of organizations are requiring certification bodies to pass a third-party review and become accredited to such standards as International Standard ISO/IEC 17024 conformity assessment—general requirements for bodies operating certification of persons (2003) and the National Commission for Certifying Agencies Standards for the Accreditation of Certification Bodies (2003, discussed further in chapters 4 and 7). In chapter 7, Sharon Goldsmith and Michael Rosenfeld investigate the possibility of accreditation of testing vendors in the certification and licensure field. Their conclusion, based partly on vendor interviews, is that a mandatory accreditation is not warranted at this time. And Roorda believes that the testing industry in the United States would be "vehemently" opposed to some type of a review system linked to standards.

The issue of whether consequences related to standards are necessary or useful for improving quality in the testing industry is open to debate, but it is clear that best practice testing organizations are using standards, internal audits, and external audits to improve quality and identify ways of continuous quality improvement.

Finally, most organizations that have a mature continuous improvement process have metrics connected to their internal process standards. Brauer (chapter 5) discusses briefly how his organization has identified metrics.

Planning and Design Activities

The second leg of the quality triangle describes the *planning and design* activities that are needed to ensure the quality of testing. Standards describe the accepted practices for a business or industry—they provide a baseline for what to plan, but they do not ensure effective implementation. Planning and design are proactive activities, that is, they look at ways to improve quality by *preventing* errors, rather than by *correcting* them after they are made.

Design refers to the systematic activities that are needed to create processes that successfully meet customer and business requirements (Ramaswamy, 1996). For example, in chapter 1 the steps in the test creation process were illustrated at a very high level. In many organizations, the details of how to create a test just developed over time. When a problem occurs (like a typo introduced when the question is entered into the item bank) a quality control step is added to catch the error before it goes further. The process is often never reviewed and designed as a whole. Is there a way to design the test creation process to prevent the introduction of the typo in the first place? In chapter 9, Rohit Ramaswamy describes a process and specific tools and steps to help focus the design of a process like test creation. Designing effective and error-free processes involves a clear understanding of customer and business requirements, the identification and quantification of key areas of risk, the use of creative techniques to identify innovative and robust solutions to mitigate risk and to meet customer requirements, and predictive methods to test whether the proposed designs will function effectively. Good design principles ensure that adequate attention is being paid to eliminating or reducing errors.

But good design alone is not sufficient; a design must be implemented and planning and project management are tools to support effective implementation. In a process organization, a process is implemented across functions or departments. Good planning and project management help assure that all the participants in a process implementation understand their work and how it relates to other parts of the project.

Several authors in this book describe challenges that they have encountered with large-scale transitions in their organizations and the value of adequate planning and project management in ensuring that these transitions take place in a systematic and controllable way.

In chapter 10, Dr. Charles DePascale describes the challenges of transitioning between test contractors, and the need for better planning to ensure smooth transitions that occur without interruption to the testing schedule, without impacting established content and performance standards, and that are invisible to test takers. Recent changes in the characteristics, use, and ownership of state assessment programs have created a different level of complexity in contractor state relationships, and the need for better planning in designing contractor transitions.

For example, in the past, changes in state assessment programs often resulted in a new start, with a break in results and no direct link between the old and new assessments. Results were not directly comparable across years and stable and steady scores were not expected at the state level. After 2001, however, this has changed. Under the NCLB act, schools are now obligated to measure and report on annual yearly progress (AYP), defined in terms of annual changes in assessment results. This has made it necessary to ensure that results are comparable across years.

Another change that has taken place recently is in the relationship between the states and the contractors. Although in the past, contractors were seen as mere as vendors of tests, today they are seen more and more as contractors responsible for providing expertise in test development and administration. This has resulted in complex interrelationships between the states and test providers that are more difficult to undo and redo than in the past.

DePascale argues that, because of these changes, there is a clear need to manage a transition process as a separate project that is different from the day-to-day operations of the assessment program. In the short term, solutions such as better planning through overlapping contracts, better communication among contractors, and clear specification of requirements can help increase the probability of a successful transition. These planning and communication activities ensure that information, data, and products created by one company can be successfully reformatted and transferred to a second company. In the long term, however, these are merely stopgap approaches, because the need for reproducing and reformatting will result in errors. If error-free test creation under all operating circumstances is the goal, there is a need to leverage current technology to develop an interoperability capability that will allow contractors to directly access test items using a common standard and use them in their tests. This level of interoperability will require systemic changes in the management and operation of state assessment programs.

The need for better management of testing contractors to ensure quality is echoed from the perspective of a state in chapter 11. In this chapter, Judson Turner outlines the tools necessary to develop a successful statewide assessment program where large contracts are awarded to private sector vendors. Turner's perspective is based upon his experience with the Georgia Department of Education. In 2003, the legal division and the testing division of the Department of Education did not partner in developing and scrutinizing vendor procurement requests. Moreover, the testing division was separate from the rest of the department. As a result, there was limited legal oversight of vendor contracts, which resulted in contractual problems and security breaches in vendor selection for administering the Georgia Criterion Referenced Competency Test (CRCT) for grades 1 through 8 in 2002. This chapter outlines the following requirements for successful management of large-scale vendor contracts by state departments of education:

- Technical expertise in the testing division
- Focus on strategic planning
- Clear understanding of what needs to be procured
- Understanding of contract provisions that can encourage vendors to perform, and their costs
- Open line of communication between the state and the vendor

As evidenced by chapters 10 and 11, the changing role of state/contractor relationships is a critical area that needs to be managed carefully to ensure the quality and integrity of tests.

Another area of transition that requires attention is the current transition from paper-and-pencil to computer-based tests (CBT). As more and more tests are being converted to the computer format, there is a need to bring systematic project management techniques to support these transitions and a growing need to address the problems associated with transition in a systematic way. In chapter 12, Noel Albertson presents some practical considerations in managing a CBT implementation using project management principles, based on his experience with implementing a CBT program at the American Institute of Certified Public Accountants (AICPA).

The chapter discusses how projects need to be organized for successful implementation, especially where there are multiple developers, or where the development tasks are shared between in-house and external resources. It also describes how a project management team may be organized and the importance and role of sponsors in providing governance and oversight of the project. Finally, the chapter discusses the importance of organizational change management activities to ensure a successful transition to CBT. This transition requires a fundamental change in business models and processes, and organizations that do not reorganize their processes and staffing will have a difficult time with their transitions. Organizational change activities are rarely integrated into the work stream for CBT projects, and this is a major shortcoming that needs to be addressed. Risk management is another critical project management activity and the chapter discusses techniques of managing project risk proactively, rather than waiting for disaster to strike.

In summary, it is always better to anticipate and mitigate errors, rather than have to correct them later. Good planning and design will help testing organizations do this, and will help states feel secure that the data and processes can be managed successfully through transition between contractors, or changes from paper-and-pencil to computer-based tests. In this period of changing relationships between states and vendors, and of the increasing use of computers in the testing industry, these planning activities may make the difference between high quality testing and chaotic operations. Many testing organizations and states are just feeling their way through these ideas—the chapters in this section should help practitioners learn about the challenges and the tools to address them.

Monitoring and Continuous Improvement

The third leg of the test quality triangle is *monitoring and improvement*. Standards create a baseline for product and process quality, planning, and design help to implement the standards and to consciously design processes that minimize errors. Monitoring and continuous improvement activities ensure that we can sustain a desired level of quality and improve over time. Standards such as ISO 9001 and CMM have continuous improvement criteria as part of their requirements.

One aspect of monitoring and continuous improvement activities is referred to as *process management.* Key process management activities include the development of process performance metrics and reporting systems for process performance. In the section on standards, we mentioned that metrics are often developed as part of an internal standard against which we can evaluate the process performance. (For example, reliability of a test is an output measure that many testing organizations use.) These targets are measures of the *outputs* of the process. Output measures are reactive measures, that is, we are not able to determine whether a process is meeting standards until the process activities are complete. In order to forestall errors before they happen, it is important to develop proactive measures of performance *within* a process that can serve as indicators for poor output performance. The chapters in this section provide examples of how to develop a process management system, potential metrics, and how to use metrics to manage a process and to continually improve the process.

The discipline of process management is discussed in detail by Rohit Ramaswamy in chapter 13. Successful process management in an organization is a strategic undertaking in order to ensure that the entire range of work activities in an organization is stable and meets customer requirements on an on-going basis. Process management also provides the needed infrastructure for continuous improvement activities that ensure that testing organizations are able to meet the on-going and changing requirements of its customers efficiently and without a reduction in quality.

Chapter 13 details a practical roadmap for implementing process management in testing organizations that consists of the following seven steps: documenting key processes; identifying process output measures; developing in-process measures that link to the outputs and can serve as leading indicators for the outputs; developing a plan for on-going monitoring of processes; developing a plan for analyzing and reporting process performance data; creating a team to review reports; and taking action and embodying a continuous improvement philosophy.

The data provided by a process management system allows organizations to identify where processes are underperforming, and where process improvement efforts are necessary. In chapter 14, Dr. David Anderson describes the use of the Six Sigma process improvement methodology at ETS to reduce errors, improve efficiency, and increase customer satisfaction. The core principles of Six

Sigma described in the chapter are the cost of poor quality (by which improvement opportunities are evaluated); the voice of the customer (through which customer needs for performance are identified), and defects per million opportunities (which is the measure of the quality of the process, with Six Sigma quality corresponding to no more than 3.4 defects per million opportunities).

The standard Six Sigma improvement methodology is called DMAIC, which is an acronym for the five steps of the methodology (define, measure, analyze, improve, and control). In Define, the improvement problem is articulated, the project is scoped out, and resource and time requirements are defined; in Measure, the extent of the problem is measured; analysis to determine the cause of the problem is conducted in Analyze; solutions for improvement are generated, evaluated and selected in Improve, and the process is transitioned to process management for ongoing monitoring in Control.

Cheating and piracy of test questions is becoming more and more of a problem for the testing industry. For this reason, metrics related to cheating are currently being added to the monitoring of test processes to reduce fraud. In chapter 15, Dr. David Foster and his colleagues write about the use of monitoring of test results to detect individual instances and systematic patterns of cheating. In this chapter, the authors describe several kinds of security problems, such as answer copying, collusion, and piracy. On the test administration side, security problems can occur when testing companies do not adequately secure their IT systems, when there are no clear rules about retakes, when there are lax legal agreements between the test providers and the test supervisors, when there is inadequate funding for security, and when there are no security plans.

Statistical analysis of test response data, called data forensics, can point to cheating patterns. For example, *erasure analysis* consists of counting total answer changes and/or wrong-to-right answer changes on an answer sheet and then determining whether the counts are anomalously high. Collusion can be detected by evaluating the degree of similarity between two or more test records. Other forms of cheating activities can be detected through other kinds of statistical analysis.

Another test security issue is the problem of item exposure in CBT, where items become exposed to test takers before they are presented in the test. Chapter 16, by Ning Han and Ronald Hambleton, deals with this topic. CBT require the daily exposure of test takers to an item bank. If there are unclear retake rules, test takers can be exposed to the items multiple times in a short time interval or pirating of items may result in exposure before the testing window is closed. This chapter describes the results of a research study evaluating the effectiveness of different item exposure statistics under different exposure models and in the presence of shifting ability over time. Further testing may provide the kinds of monitoring that is needed and the statistics that need to be collected to produce accurate estimates of item exposure probability under different test and ability conditions.

A common theme of all the approaches described in the chapters in the monitoring and continuous improvement section is that they all require data for analysis. Process management and improvement is not possible unless timely data on process outputs and in-process metrics are available to the process management or to the Six Sigma team. Similarly, data forensic analysis and the estimation of models to evaluate item exposure require extensive data on test results. At this time, the data needed to make intelligent business decisions is not always available, or if it is, is in fragmented systems from which extraction is difficult and time consuming.

Chapter 17, by Richard Dobbs and Stuart Kahl, describes how customers are demanding that testing organizations be "faster, better, and cheaper" and how this is not possible without the development of integrated systems that support all the major processes associated with test development, test administration, scoring, and reporting. The complexity of customer demands and the flexibility and speed that will be needed in the future cannot be met without the definition of key process metrics, and the use of integrated enterprise systems to track progress on these metrics and to allow effective management of the work flow.

Conclusions and Lessons Learned

In conclusion, improving a testing organization requires effective integration of a whole series of process tools. These include implementing standards related to both testing and more generic processes, using tools for process design and planning, and monitoring processes and applying continuous improvement tools as needed. Leadership must be engaged in the implementation of any organization-wide initiative.

The summary of the chapters mentioned earlier begins to illustrate how all the parts of the conceptual model work together to enable an effective implementation. Take leadership for example. Many and multiple standards are available to the testing organizations—leadership is necessary to decide which of the many external standards are most relevant to the strategic plans of the organization and how internal standards should be developed and implemented. Updating documentation to keep current with standards and preparing for internal or external audits are easily postponed unless leadership determines that it is a priority. Leadership needs to reward process managers who design processes to prevent errors rather than the *fire fighters* who fix problems before the test goes to the printer or the scores are mailed to the examinees. Processes work across departments, and departmental managers need to be trained to manage work to meet process needs rather than defending their fiefdom. The very act of reserving time to plan and design a process rather than jumping in and *doing* requires a discipline that must be modeled and encouraged by leadership. Finally, the monitoring and improvement requires a management team for each process and a leadership team that reviews the overall process metrics and determines the organizational priorities for improvement and design projects.

Some of the standards that are relevant to testing organizations, (ISO/IEC 17024, ISO 9001:2000; ISO 27001 & Chrissis, Konrad, Shrum, 2003) include continuous improvement standards and support the monitoring and improvement leg of the quality triangle. Metrics are a necessary part of process management, product design, and evaluating whether an organization is meeting standards.

Chapter 18 suggests how an organization might go about implementing the test quality triangle and the future work needed to improve testing.

The chapters in this book provide a diverse perspective on how these attributes work in practice in many state governments, testing sponsors, and testing organizations, and what still needs to be done as we raise the bar on the quality of testing in the United States and abroad.

References

American Educational Research Association, the American Psychological Association, and the National Council on Measurement in Education.

(1999). *Standards for educational and psychological testing.* Washington, DC: American Educational Research Association.

Chrissis, M. B., Konrad, M., & Shrum, S. (2003). *CMMI: Guidelines for process integration and product improvement.* Boston, MA: Pearson Education, Inc.

International Organization for Standardization. (2000). *ANSI/ISO/ASQ Q9001-2000 American national standard: Quality management systems—requirements.* Milwaukee, WI: American Society for Quality.

International Organization for Standardization. (2003). *International standard ISO/IEC 17024 conformity assessment—general requirements for bodies operating certification of persons.* Geneva, Switzerland: Author.

International Organization for Standardization. (2005). *International standard ISO/IEC 27001 information technology—security techniques—information security management systems—requirements (2005-10-15).* Geneva, Switzerland: Author.

Juran, J. M. (1999). How to think about quality. In J. M. Juran & A. B. Godfrey (Eds.), *Juran's quality handbook* (5th ed.). (pp. 2.1–2.18). New York: McGraw-Hill.

Pande, P. S., Neuman, R. P., & Cavanagh, R. R. (2000). *The Six Sigma way: How GE, Motorola and other top companies are honing their performance.* New York: McGraw-Hill.

Ramaswamy, R. (1996). *Design and management of service processes.* Prentice-Hall.

Smith, D., & Blakeslee, J. (2002). *Strategic Six Sigma: Best practices from the executive suite.* Hoboken, NJ: John Wiley & Sons.

Part II

Leadership

3

Leadership in Testing

Richard C. Smith
PricewaterhouseCoopers, LLP

Introduction

In the previous chapter, Dr. Rohit Ramaswamy described the *quality triangle*, the conceptual framework used to organize this book. This triangle is shown in Figure 3.1. The three legs of the triangle are *implementing standards* for quality, employing the appropriate *design and planning* tools to prevent errors or build in quality, and *monitoring and improvement* to make sure that process performance is sustained and improved over time. At the center of the triangle is *leadership engagement*, which is the glue that binds these different aspects of the triangle together. This chapter focuses on the importance of leadership in ensuring the successful implementation of the model in the testing industry. *Successful implementation of this model is dependent on the energy, focus, and commitment of leadership. Not just top management, but all leaders in the organization supporting the improvement initiative.*

Much has been written about the failures of quality management programs and process improvement initiatives in the past decade. Researchers and practitioners who have studied this extensively find that there is one primary cause for this failure, and that is that the leaders still don't get it. We continue to witness less than satisfactory results from major enterprise-wide efforts. In fact, the recent research survey (PricewaterhouseCoopers & Opinion Research, 1998) supports the statement that 70% of quality/process initiatives don't achieve the goals that senior management expects. Therefore, it is important to identify the leadership behaviors that are really needed to drive Quality Management, Business Process Management, Six Sigma, and other process-centered initiatives to significantly improve business results.

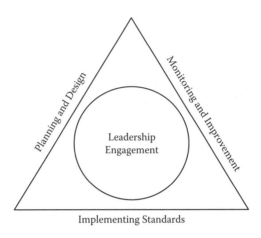

Figure 3.1 The quality triangle.

Critical Success Factors for Successful Improvement Initiatives

Let's look at the critical success factors (CSFs) for change and discuss several points of view regarding the leadership behaviors that appear to be necessary if not mandatory for a successful deployment of Six Sigma and other improvement initiatives. All major quality improvements in organizations result in two types of change—changes in the work (often involving technology) and changes in the attitudes and behavior of people. Change management is the term often applied to the tools used by leaders to encourage changes in attitudes and behavior of people, changes that are necessary to ensure the desired changes in how work is done are actually implemented.

Two research reports identify and validate the required CSFs (Conference Board, 2005; PricewaterhouseCoopers & Opinion Research, 1998). The Conference Board research asked senior management of various companies to identify their major change drivers and the factors that supported and impeded successful management of change. When asked to identify the drivers that prompt companies to develop change capacity, the management group selected improving organizational performance and cost reduction as the major drivers of change. The most common change initiatives underway included business process improvements, organizational restructuring, and change in senior leadership.

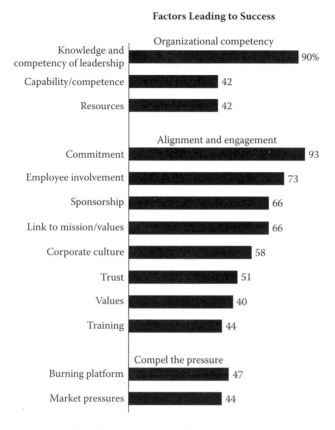

Figure 3.2 Factors leading to successful change.

Figure 3.2 shows the main leadership factors contributing to successful change. The number one factor, commitment, is critical to achieving alignment and engagement across the organization. Figure 3.3 shows the top challenges to effecting change. The top challenges are people issues, organizational resistance, and communication weaknesses.

The findings also indicate that initiating change starts with the CEO and other senior officials. Significant organizational change such as process improvement requires more hands-on leadership from top brass; however, sustaining change requires day-to-day management at the unit leader level. The role of middle management is also critical because their position in the organization allows them

**Top Challenges
to Change Management**

Organization

Organizational resistance — 58%
Strategic alignment — 32
Downsizing — 25
Structure — 23
Lack of defined mission — 18
Outsourcing — 13

People

People issues — 59
Employee resistance — 39
Lack of trust — 34

Process

Communication weaknesses — 47

Figure 3.3 Challenges to successful change.

to act as facilitators or road bocks to successful implementation. Middle managers are ultimately responsible for managing the resources that will participate in the change initiatives. As with any initiative, effectiveness is greatest when it is driven from the level closest to the people who are actually affected by it.

When key corporate, middle, and local leaders are seen as sponsors and embracing the goals of change, it encourages alignment both at the organizational and departmental levels. Sponsors should communicate the compelling reason why change needs to occur.

These findings from the Conference Board study are mirrored in the findings from the PricewaterhouseCoopers study (1998). The study of 500 companies identified the top four success factors of transformation change initiatives as top management sponsorship (82%), treating people fairly (82%), involving employees (75%), and communications (70%). The top four barriers are competing resources (48%), functional boundaries (44%), change skills (43%), and middle management resistance (38%).

The key messages from the Conference Board and the PricewaterhouseCoopers surveys are that all change management success factors and 5 out of 10 barriers to change are people related, and that without senior management sponsorship and commitment, organizational change cannot take place. Because quality improvements inevitably involve change, the vital role that sustained leadership support plays in ensuring the successful deployment of quality programs cannot be minimized.

In our experience in the testing industry, the factors required for supporting change are the same as in other industries and the mistakes made by leaders in the testing industry are similar to those made by leaders in other industries. We have often seen senior management in the testing industry underestimate the requirements of sponsorship. They assume that sponsorship can be accomplished by simply delegating the responsibility to a middle manager and requesting monthly reports. Many testing organizations have a quasi-academic culture, a culture that expects lengthy discussion and some academic license to do the work in the manner each person sees as best. And because some of the work is knowledge work that is hard to document, some academic expertise is indeed required to obtain a quality product. This culture in testing firms leads senior management to postpone or avoid major decisions, especially if staff may disagree or feel they have not been adequately consulted.

In one leading testing organization, we conducted a survey of staff and found that communication from senior management, to middle managers, and then to front-line supervisors was not working. Staff at all levels felt that the objectives of the process improvement projects were not clearly laid out for them and that the end game of the project was not clear to them. The staff was not involved enough in many of the changes and not willing to accept the changes suggested.

In chapter 14 of this book, Dr. David Anderson of the Educational Testing Service (ETS) writes about the importance of senior management engagement in their process improvement efforts. Anderson recommends that senior management be trained in process improvement methodology and in their roles of providing top-down oversight and commitment to subsequent groups before anyone else in the organization. This helps senior managers to facilitate the deployment in their organizations. Without this facilitation, there will be a lot of organizational resistance to the changes brought about by the initiative.

Dealing with resistance is not easy for any organization, and in our experience, organizations that have individuals with high levels of education and significant professional expertise, including testing and professional service companies, tend to use passive resistance as a way to block or subvert progress. In the following section, we look at what the literature says about resistance and its causes.

Overcoming Resistance

An unpublished internal employee survey conducted by PricewaterhouseCoopers showed that employees resist change because they fall into one of three categories. The majority of resistance results from staff not knowing what the change or improvement initiative is all about. The staff does not really understand the goals and objectives of the initiative. Perhaps the goals and objectives haven't been communicated, or they haven't been communicated in a way that employees can understand, the "WIFM" (what's in it for me?). This group typically represents 50% to 60% of the employee population. The solution to overcoming resistance for this group is more open and effective communication. We deal with communication issues in a later section.

The second category of staff is unable to make the transition. This is because they have not been trained or do not possess the skills necessary to work in the new environment. This category typically represents 30% to 40% of the employee population. Therefore, if an organization's quality initiative requires new technical, process, or technology skills then the organization must educate and train the staff to prepare for this transition. For example, in one testing organization, we found front-line supervisors were expected to lead the transition to a new process but were not trained in how to communicate about the change initiative. A training program was initiated that explained the process changes, the reasons for the changes, the role expectations for supervisors, and how to communicate the changes. Having knowledge about the "how to" of the process change significantly increased the confidence of front-line managers and their willingness to support the change.

The third category consists of those who are unwilling to change. This reflects a population who, for whatever reason, will not support the quality program. Senior management has two choices in this situation: either ignore the group or remove it from the organization. In either case, senior management must realize that this group will create some level of disruption and be prepared to deal with it.

We have talked about the importance of senior managers and the resistance they may face in implementing improvement initiatives. But what should they do to demonstrate commitment and to overcome resistance? In the following section, we examine the role that leaders play in organizations.

The Role of Leaders

To understand leaders' roles, let's reflect on the teachings of leadership guru John Kotter. Kotter (2001) outlines several key themes that leaders must apply use to successfully lead change initiatives in their respective organizations. Dr. Kotter's points can be summarized as follows:

- Set direction and vision for change
- Align people versus organizing staff
- Motivate and inspire people
- Reward and recognize success
- Communicate, communicate, communicate

Kotter believes that being able to articulate a clear vision and strategy for change is one of the most crucial behaviors of a transformational leader when introducing any kind of change initiative into the organization. Vision must include a conceptual roadmap or set of blueprints for the future organization.

The importance of vision and strategy was also validated by Dr. Warner Burke, professor of organizational psychology at Columbia University. He stated at his speech at the 2004 Organization Development Network Conference: "So, as we know, it's all about direction and values, and the why regarding the necessity for change." Burke further stated: "We also know that it is important to distinguish between leadership and management. There is overlap of course, but change requires leadership far more than management, particularly at the onset. Leadership is also required to sustain the change. The wise leader will be very participative and involved in any implementation." Finally, Burke stated, "We know how critical it is for leaders, especially in time of significant change in the organization, to openly and clearly match their words to their actions. The audio needs to match the video."

In addition to vision and strategy, Kotter outlines the importance of "aligning people" to support the change. Kotter argues that getting staff aligned to the desired change is a leadership role. This will involve communicating to the target population about the change, trying to help people to comprehend a vision about an alternative future that is better than the current state, and empowering people to make the personal change necessary to be successful. These were corroborated in the findings from the two research surveys mentioned earlier in this chapter.

As we reflect upon our consulting engagements in the testing industry, we find that functional barriers and not empowering staff are two leading factors that inhibit successful quality or process improvement initiatives. For example, some testing organizations have strong functional departments like test development, statistical analysis, information technology, and client management. Managers may not encourage or even actively discourage cross-functional thinking among departments. Teams with staff from test design, client management, and analysis working collaboratively to streamline a process is often not comfortable for the testing organizations because it threatens the departmental authority and decision-making power of the middle managers. In order to conduct a significant or *breakthrough* process improvement, we must engage cross-functional teams and allow them to break old paradigms about how work gets done efficiently. Therefore, leadership must step forward and help staff cope with the change in direction and provide the inspiration for staff to work in the new process world. Kotter views these teams as *coalitions* that understand the vision and work at achieving it.

For testing organizations to be successful in an increasingly competitive environment, more assertive leadership will be required. For approaches on this, the ideas and concepts of leadership guru Noel Tichy are worth noting. In his book *The Leadership Engine* (Tichy, 2002), he notes that driving quality programs, such as Six Sigma, requires leaders to possess *edge*. Tichy defines edge as a top leader's ability to make the tough decisions affecting the long-term success of a business. Such leaders challenge and recommend unpopular or unusual ideas as part of focusing the organization on needed changes.

Tichy further points out that leaders should bring positive energy to the organization and have *ideas* about the business that drive performance improvement. Tichy advocates the use of an executive leadership workshop as a vehicle for engaging a critical mass of managers and building the momentum for transition. The workshop format has proven to be very successful for organizations like GE, Honeywell, Ford Motors, Wellmark, and others. In this workshop leaders at all levels receive feedback on their leadership style, develop teamwork aimed at transforming the organization, identify processes for improvement, and develop a game plan to mobilize staff through the transition.

As we look at the leaders and managers in the testing industry, it is interesting to reflect on how many of these leaders really possess the edge factor that Tichy feels is so important to organizational transition. How many leaders clearly demonstrate that they have positive energy to change the organization and how many leaders have articulated great ideas that would drive performance improvement in their respective testing business? Many of the leaders that Tichy uses as examples have been champions of quality or process initiatives in their respective firms, such as Jack Welch of GE, Larry Bossidy of Allied Signals, Dan Burnham of Raytheon, or Ellen Gaucher of Wellmark. Few of the testing leaders of today have a background in quality or process initiatives and this lack of expertise may inhibit the capability of the industry as a whole to embrace the leadership discipline necessary to lead change initiatives.

Making Quality Happen—The Raytheon Example

In recent years, many of the examples of strong leadership commitment and alignment have come from companies that have implemented Six Sigma. Whether you are an advocate for the Six Sigma approach or of some other approach is not the issue. The companies who have totally committed their energy and resources to the implementation of quality improvement programs have seen significant positive financial, customer, and process improvements, and positive cultural change. Those companies include Dow Chemical, Bank of America, Merrill Lynch, Caterpillar, and Wellmark. In every one of these cases, success has been achieved because of the total organizational commitment to the effort and the complete engagement of the chief executive.

One excellent example of this type of leadership is Dan Burnham, former CEO of Raytheon, who led Raytheon through major organizational transition. As CEO he recognized that the journey of changing the organizational landscape at Raytheon began with him. He clearly recognized that this initiative was more about changing the culture than about implementing a quality program.

During an interview that was conducted for the book *Strategic Six Sigma* (Smith & Blakeslee, 2002) Burnham stated:

> If you think about Six Sigma as another quality program, then it deserves as much intensity as all other initiatives that can go into a big company. But to the degree that you see Six Sigma as a culture changer—something that will profoundly affect the organization—then by definition, it takes the passion and obsession of the CEO to make it happen. At Raytheon, we saw Six Sigma as a way to profoundly change our culture, and therefore it started with me and ends with me. I include language on it at almost every meeting that I have, to the extent that people's lips almost move in synch with mine on this subject.*

Burnham is an excellent example of a successful leader who has mastered another aspect of Tichy's leadership model that "good leaders are good teachers." In his book, Tichy states that good leaders have points of view that they can clearly articulate and convey to others. This ability to teach can encourage a change in behaviors. Certainly Burnham was able to do that as he transformed Raytheon from a *behemoth defense company* to a competitive company with strong process and customer focus.

In the previous section and in this one, we have talked about vision, edge, and passion. In the following section, we focus on another critical leadership role that is no less important, and that is *communication*.

The Role of Communication

All the research studies and experts we have mentioned so far point to the need for effective communications. Communication is fundamental to an organization's change efforts.

The vital importance of communications has been repeatedly stressed. In order to succeed, communication activities must be reciprocal; management has to listen as well as talk. The change management study from the Conference Board (2005) mentioned

---

* Communication with author.

previously highlights the top techniques for successful communications. Among the key techniques are:

- Clarifying the rationale for change
- Continuous communication/repetition
- Championing the change
- Two-way communication
- Explaining how change is achieved
- Using multiple communication methods
- Inviting employees to participate

Companies are using a variety of techniques for fostering their respective quality initiatives. A recent company that clearly understood the value of communication was Dow Chemical in its roll-out of the Six Sigma program.

As detailed in the book *Strategic Six Sigma* (Smith & Blakeslee, 2002), a Dow executive, Tom Gurd, explains, "we are very focused on creating constancy of purpose for Six Sigma. Consequently our leaders have maintained a focused, high energy approach to communicating the urgency of Six Sigma to the company employees around the world" (p. 253).

Gurd further states: "a company's leaders must never let up on communicating Six Sigma priorities. They must consistently display visible and vigorous leadership, so that people have no doubt of where the organization is headed" (p. 253).

Our change management experts have also emphasized the critical need for effective communications. Dr. Warner Burke states: "gaining commitment from organizational members to the future state, a plan is critical; gaining commitment to implementing the plan is even more critical. Therefore, leadership needs to make certain that the plan is adequately communicated and generate the energy to support the transition" (Smith & Blakeslee, 2002, p. 152). Burke stresses that communications with organization members must answer the question "How will this change affect me?" and respond to the question "What's in it for me?"

In our work with process improvements or quality programs in the testing industry, we have repeatedly promoted with senior executives the need to develop extensive communication programs to support the performance improvement initiative. We have coached these executives that clarifying the change, answering relevant questions, identifying resistance, and building trust with staff is an important competency for implementing quality programs. In most initiatives we have dedicated a small team of staff to the *communications change team*. These team members are not communication experts, but rather respected employees whose opinions are valued by the others and who understand the communications needs of staff.

A successful approach to communication, called *cascading communications*, was successfully adopted by ETS in 1995–1996 to facilitate the redesign of the test creation processes. ETS embarked on a cascading communications effort that was designed to engage managers and staff through a series of work group meetings or *labs* to introduce the different aspects of the new test creation process. This effort is described in an article in the December 1996 edition of *HR Magazine* (Wild, Horney, & Koonce, 1996). As indicated in this article, this approach not only encouraged involvement in the redesign process but also set the stage for buy-in and ownership for the new process.

In summary, we must again emphasize that effective communication is critical to creating a motivating vision for the organization—this is a competency leaders in the testing industry must acquire.

Leadership Role in Sustaining Change

So far, we have talked about a leader's role in creating a vision for change and in initiating change within an organization. But just leading the change process alone is not sufficient; once the changes are in place, leaders must sustain their commitment so that the improvements that have been put in place achieve their full benefits over time.

Smith and Blakeslee (2002) point out that to sustain major Six Sigma initiatives the leadership must be totally involved and drive the initiative. After investigating more than 20 Six Sigma implementations, the authors found that to sustain the implementation over time certain critical success factors were necessary. The graph in Figure 3.4 depicts the impact of sustaining leadership over time on the dollar savings resulting from a Six Sigma company-wide implementation. Curve A represents organizations where the leadership commitment continued through four years. As the length of the sustained commitment of senior management declines, the process improvement program savings also declined, illustrated by lines C and F.

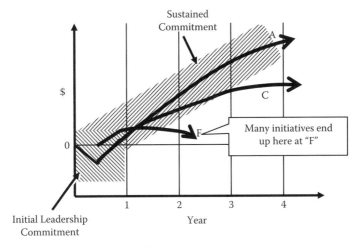

Figure 3.4 Sustaining leadership over time.

The key factors that differentiate an organization that moves along curve A from one that moves along curve F are:

- Business integrated with culture
- Focus on customer value
- Project selection linked to strategies
- Ability to change culture and behaviors
- Leadership and alignment of managers
- Metrics and tracking

All the previously listed factors require active leadership involvement throughout the process improvement implementation.

Conclusions and Lessons Learned

So, in the end, what does all this mean? After reviewing the leaders who have successfully deployed a major quality or process initiative and looking at all the research, we have collected several best practices that represent what senior leaders really need to focus on. We describe these and the lessons learned in the rest of this chapter.

Leadership Best Practices

1. The CEO must effectively articulate the compelling reasons for the organization to embrace quality. These are not responsibilities that top business leaders delegate to anyone else. Leaders must set the overall strategy. A sense of urgency is required. We noted Dan Burnham of Raytheon earlier in this chapter as a great example of this behavior.
2. These messages must be effectively communicated at all levels of the organization in order to build a broad base of support. Leaders who don't communicate frequently or build a strong *coalition* of managers in the organization are likely to fail or not meet the goals for the business process management or Six Sigma initiative.
3. Linking improvement projects to the corporation's strategic goals is essential to maintain the buy-in of executives who own both the resources and the budgets for major initiatives. The more important the project is to meeting corporate strategic goals, the more likely it is that executives will support the projects. This view was clearly articulated by the chairman of Dow Chemical, who positioned Six Sigma as a strategic program. Dow implemented Six Sigma to drive three key business goals: (a) customer loyalty and growth, (b) productivity enhancement, and (c) culture and people alignment.
4. Generating short-term project wins to build organizational momentum is essential to gaining longer-term acceptance for the vision and direction within the organization. Because most quality initiatives represent radical change, resistance is likely until people see the personal and organizational benefits. The companies who have been most successful at implementing quality programs, such as Dow Chemical, Caterpillar, and Bank of America, have done so by obtaining early results that drive confidence in the approach and sustain momentum.
5. To sustain success, a company's top leaders must effectively manage both the hard (performance expectations and metrics) and soft (people and culture) sides of organizational change. Embarking on significant process change requires culture change, which usually includes dealing with metrics, process, and people.
6. Implementation must be linked to compensation and other human resource practices. Doing this early in the initiative will accelerate the transition. In the PricewaterhouseCoopers research study (1998) presented in this chapter, executives interviewed felt they could have accelerated the speed of their initiatives if they had tied compensation and incentives directly to the change objectives at the start for the initiative.

7. Another critical success factor is training and skills development of the staff who are leading by the process improvement efforts. One of the key differentiators of Six Sigma is the heavy emphasis on Green Belt and Black Belt training. But the fact is that Six Sigma training builds skills and capabilities for the organization. At ETS during the re-engineering of the test creation process, they identified the technical skills that staff needed to work in the new high-tech environment. They developed and delivered the training to support the transition to the new item banking software so staff could work successfully. More recently ETS has migrated to Six Sigma as their process improvement methodology. (See chapter 14 for information about Six Sigma at ETS.)

8. Focus on results by creating clear expectations and accountability at all levels of the organization. Again, clear communication, planning, establishing metrics, and reward systems that are linked to the strategic objectives of the process improvement program. Here it is essential that leaders select and prioritize projects that will drive the biggest value or customers and employees and generate financial results for the organization.

We have observed positive evidence in our work with testing industry management teams that the importance of the previously mentioned factors is clearly understood. Sustained focus, however, is difficult and few leaders have yet developed a mature quality management culture in their organizations.

The focus of this chapter has been to highlight the leadership elements that drive performance improvement through a quality or process improvement implementation. It is not easy—while the leader is leading the change initiative he or she is also required to deal with all strategic directions of the organization, with financial planning, and the myriad other top priority issues for the organization. The effective leader must balance the priorities, understanding the lessons learned and balancing them with the other organizational needs.

The key leadership factors for achieving successful change in the testing industry are no different than in other industries. What will be different is how you design and implement the initiative to work in your culture. If you as a leader are having trouble balancing priorities and leading the change effort, you can bet that the rest of your organization is having priority issues and will be following your lead. This is why it is no mistake that leadership is in the center of the quality triangle.

References

Burke, W. W. (2002). *Organizational change: Theory and practice.* Thousand Oaks, CA: Sage Publications.

Burke, W. W. (2004). *Organization development.* Paper presented at the OD Network Annual Conference.

Conference Board. (2005). *Effecting change in business enterprises: Current trends in change management.* (Conference Board Research Report.) New York: Author.

Kotter, J. (2001, December). What leaders really do. *Harvard Business Review*, 3–11.

PricewaterhouseCoopers & Opinion Research (1998). *Accelerating transformational change.* Washington, DC: PricewaterhouseCoopers.

Smith, D., & Blakeslee, J. (2002). *Strategic Six Sigma: Best practices from the executive suite.* Hoboken, NJ: John Wiley & Sons.

Tichy, N. M., & Cohen, E. (2002). *The leadership engine.* New York: Harper Collins.

Tichy, N. M., & Devanna, M. A. (1990). *The transformational leader.* New York: John Wiley & Sons.

Wild, C. L., Horney, N., & Koonce, R. (1996, December). Cascading communications creates momentum for change. *HR Magazine*, 94–100.

Part III

Standards

4

Standards in the Testing Industry

Cheryl L. Wild
Wild & Associates, Inc.

Joan E. Knapp
Knapp & Associates International, Inc.

Introduction

Tests are designed to measure a variety of constructs such as cognitive abilities, psychomotor skills, knowledge, verbal and mathematical ability, and professional competence. Testing is an integral part of our lives, from the dreaded content examinations required by the No Child Left Behind mandate for grades 3 through 8, to those feared pop quizzes in junior high school, to the ominous high school exit examinations, to challenging college admissions tests like SAT or ACT that help determine where you will spend four years of your life, to formidable midterms that indicate to parents what they might expect of their children at the end of the first semester of freshman year, to arduous graduate or professional school admissions tests such as the GRE or the MCAT, and, finally, grueling licensure and certification examinations after professional or occupational education and training has been completed.

Most test results and their interpretation can have high-stakes consequences for those who are being tested. Such consequences on the downside can include failure to graduate from high school, not being admitted to the college of one's choice, or not obtaining an occupational license and thus not being able to work in one's vocation. Test takers are not the only ones subject to the consequences of testing. As a result of No Child Left Behind legislation, schools and districts face sanctions if their students do not meet performance targets. Accreditation standards for organizations, particularly in healthcare, are sometimes based on having a minimum percentage of certified professionals on staff—if employees cannot past the tests, organizations may not be accredited. Because test results have critical consequences to test takers and other stakeholders, it is crucial that tests be fair, reliable, and valid for their intended purposes.

Fair, valid, and reliable are terms that have specific *psychometric* meaning. Consumers of tests, however, have needs that go beyond psychometric attributes. Test takers' expectations include speedy receipt of score reports; friendly and helpful customer service, whether it be related to ordering, shipping, test administration, or questioning score reports; confidential and secure handling of information about test questions and test takers; and simple and user-friendly online services. The testing industry must meet the needs of its clients. Like any other company, a testing company, if it is to be successful, must meet the expectations of its customers and conduct its business in an effective and efficient manner. Costs must be contained, management practices should promote continuous improvement, and ethical standards related to business practices must be upheld.

Fortunately for consumers of tests, there are standards that can be applied to both testing products as well as appropriate business processes. Standards are tools that many industries use to define expectations for industry products, services, and processes. This chapter defines standards, suggests why standards are important, and identifies the most frequently used standards. Industry uses of standards, including the use for accreditation of testing products and processes, are described. Finally, suggestions for how testing companies and testing sponsors can use standards to increase product and process quality are presented.

We learned about these standards, their benefits, and their drawbacks from Web searches and discussions with representatives of standards organizations and through interviews with testing companies and test sponsors. In all, twelve interviews were conducted: seven with test vendors and five with test sponsors. The vendors covered a number of testing fields, including certification/licensure (3), education (3), and industrial/organizational (1). Eight of the twelve use both paper-and-pencil and computer-based delivery, two deliver tests primarily in a paper-and-pencil mode, and two deliver tests primarily via computer. Four test primarily in the United States whereas eight test internationally.

What Are Standards?

ISO/IEC Guide 2 (1996) notes that a standard is a

> document established by consensus and approved by a recognized body that provides for common and repeated use, rules guidelines or characteristics for activities or their results aimed at achieving the optimum degree or order. (p. 10, under clause 3.2)

Industry standards and guidelines are typically forged from a consensus among a substantial number of people who are knowledgeable about the processes or products that the standards address. Additionally, opportunity for public comment is included in the development process. The ultimate goal of a standard, or set of standards, is to bring about standardization in the processes and products that may emerge from these processes. In addition to standards being consensus driven, the American National Standards Institute (ANSI) advises that the standards-development process should be characterized as having due process, openness, and balance.

Why Are Standards Important?

As noted previously, stakeholder groups in the testing industry have a lot to gain or lose as a result of the scores produced from the testing process. Test takers perhaps have the most at stake because test scores can deeply affect their life choices. If the tests are flawed or the scores are not accurate, these choices may be inappropriately influenced. Poor-quality tests and scoring errors can seriously impact the reputation and bottom line of a testing company and provide the fodder for its competitors to win lucrative contracts. Test sponsors such as certification organizations and departments of education whose names are associated with poor-quality tests can suffer losses in credibility and even financial penalties when problems become public.

Standards provide guidance to testing companies, test sponsors, and test consumers about what constitutes minimally acceptable procedures for developing and delivering tests, managing a testing organization, maintaining security, and providing quality management and continuous improvement. Adhering to relevant standards can raise the quality of testing and testing processes.

What Standards Are Relevant to Test Makers and Test Takers?

Standards related to the making and taking of tests come in many types and forms. For purposes of discussion, we have organized them into three groups: standards that focus specifically on testing products and processes and are intended as guides (Table 4.1); standards that are used to accredit programs that have an examination component (Table 4.2); and standards that are not specific to testing but that relate to systems, processes, and so forth that are used in the production and delivery of tests (Table 4.3).

The relationship of these three sets of standards is illustrated in Figure 4.1. The criteria that compose the standards in each of the three tables are represented by a circle. The content of the standards in Tables 4.1 and 4.2 overlap somewhat in terms of test design, development, and delivery standards. The standards in Table 4.3 are generic—they do not contain specific guidance concerning testing—thus there is no content overlap in the circles representing Tables 4.1 and 4.3. There is some content overlap, however, between Tables 4.2 and 4.3, because some of the standards in each table include organizational requirements and quality management and/or continuous improvement requirements.

Standards Specifically Related to Testing

The thirteen standards listed in Table 4.1 relate specifically to testing. The most comprehensive and well known are the *Standards for Educational and Psychological Tests*. They have become the foundation and sometimes the template for a number of testing standards. These standards are recognized not only in the United States, but also by some psychological associations in Europe. The European Federation of Psychologists' Associations (EFPA) has developed an evaluation model for psychological tests. Although the EFPA does not conduct reviews/audits of tests and programs per se, some of the

TABLE 4.1 Testing Standards with No Accreditation Component

Standards/Guidelines	Description
Code of Fair Testing Practices in Education (American Educational Research Association, American Psychological Association & National Council of Measurement in Education, 2004)	• Purpose is to guide professionals who develop or use tests in an educational context (e.g., admissions, assessment, and student placement) • Addresses developing and selecting tests, administering and scoring tests, reporting and interpreting test results, and informing test takers • Intended to be consistent with the Joint Technical Standards mentioned later and to safeguard the rights of test takers
Development, Administration, Scoring, and Reporting of Credentialing Examinations: Recommendations for Board Members (Council on Licensure, Enforcement, and Regulation [CLEAR], 2004)	• Purpose is to provide guidance to licensing boards in appropriate and acceptable practices in test development, test administration; statistical analysis and research, scoring and reporting; additional sections focus on computer-based testing and examination security • Each section ends with questions to ask test vendors or state departments offering testing services to state licensing boards or regulators
EFPA Review Model for the Description and Evaluation of Psychological Tests: Test Review Form and Notes for Reviewers (European Federation of Psychologists' Associations, 2005)	• Purpose is to provide a structure and tools for the review and evaluation of psychological testing instruments • Contains sections for providing qualitative descriptions (of the instrument; content/use classification; measurement and scoring; computer generated report; supply, conditions and costs); sections for a quantitative evaluation of the instrument (test materials, norms, reliability, validity and the quality of the computer-generated reports), and an overall instrument evaluation • Modeled on the form and content of the British Psychological Society's (BPS) test review criteria and criteria developed by the Committee of Test Affairs (COTRAN) of the Dutch Association of Psychologists (NIP)

(Continued)

TABLE 4.1 Testing Standards with No Accreditation Component (Continued)

Standards/Guidelines	Description
ETS Standards for Quality and Fairness (2000)	• Purpose is to provide standards for internal audits of ETS programs, the results of which are reviewed by an external committee • Includes developmental procedures; suitability for use; customer service; fairness; uses and protection of information; validity; assessment development; reliability; cut scores, scaling and equating; assessment administration; reporting assessment results; assessment use; and test takers' rights and responsibilities. Closely reflecting the Standards for Educational and Psychological Tests
Guidelines for Computer-Based Testing (Association of Test Publishers, 2004)	• Purpose is to supplement the Standards for Educational and Psychological Tests (1999) as the standards apply to computer-based tests • Presents CBT guidelines in the areas of planning and design, test development, test administration, scoring and reporting, psychometric analysis, stakeholder communications, and security
IMS Question and Test Interoperability Overview: Version 2.1 Public Draft Specification (IMS Global Learning Consortium, Inc., 2006)	• Purpose is to provide standards for software development that will enable the exchange of item and test results data between authoring tools, item banks, test construction tools, learning systems, and assessment delivery systems • Presents an abstract data model for design of question and test programs with specific standards for interchange between systems • Focuses on the distributed learning environment and outlines specifications on how software must be built to meet the test requirements
International Guidelines on Computer-Based and Internet-Delivered Testing (International Test Commission, 2005)	• Purpose is to raise awareness and highlight what constitutes good practice among all stakeholders in computer-based and Internet testing in a global context • Divided into four areas: technological issues in computer-based and Internet testing; quality issues in CBT and Internet testing; control over CBT and Internet testing; security and safeguarding privacy in CBT and Internet testing • Guidelines under each area present responsibilities for test developer, test publishers, and test users • Intended to supplement the ITC Guidelines on Test Use (2000)

ITC Guidelines on Adapting Tests (Hambleton, 2005)	• Purpose is to provide practitioners with a single source for advice on test adaptation in different languages • Covers four areas: context, test development and adaptation, administration, and documentation/score interpretations
ITC Guidelines on Test Use, Version 2000 (International Test Commission, 2001)	• Purpose is to produce a set of guidelines relating to the competencies required of test takers and to cover all areas of test use • Intended to apply to test users and is based on the following supposition: "A competent test user will use tests appropriately, professionally, and in an ethical manner" • Presented under two main headings: take responsibility for ethical test use and follow good practice in the use of tests with subheadings under each that help the user define the competencies and skills required to carry out the testing process
Principles for the Validation and Use of Personnel Selection Procedures (Society for Industrial and Organizational Psychology, 2003)	• Purpose is to provide guidance to professionals who are involved in the validation and use of selection procedures • Topics include sources of validity evidence, generalizing validity evidence, and fairness and bias. The document does not mandate specific procedures or provide advice on how to comply with applicable local, state, or federal laws
Responsibilities of Users of Standardized Tests (Association for Assessment in Counseling and Education, 2003)	• Purpose is to promote appropriate test use for professionals in the counseling and education community • Contains guidance on seven topics: qualifications of test users, technical knowledge, test selection, test administration, test scoring, interpreting test results, and communicating test results
Rights & Responsibilities of Test Takers: Guidelines and Expectations (Joint Committee on Testing Practices, 2000)	• Purpose is to inform and educate not only test takers, but others involved in the testing enterprise so that testing measures can be used validly and appropriately • Identifies ten rights and responsibilities of test takers and ten responsibilities of testing professionals
Standards for Educational and Psychological Testing (American Educational Research Association, American Psychological Association & National Council of Measurement in Education, 1999)	• Purpose is to address professional and technical issues in test development and the appropriate use of tests in education, psychology, and employment • Mainly used as a reference or benchmark in a variety of contexts (e.g., legal, educational, policymaking) • The standards are the foundation for the technical requirements for accreditation by a number of organizations such as ABNS, NCCA, ABMS, and others

TABLE 4.2 Standards Developed for the Purpose of Accrediting Testing Programs

Standard	Description	Type of Audit	Number of Programs Recognized
American Board of Nursing Specialties Accreditation Standards (American Board of Nursing Specialties, 2004)	• Purpose is to protect the public interest by applying specific standards to the quality of specialty nursing certification programs • Apply to certification tests only, not the certification programs, in general • Include technical requirements for test validation, test development, reliability, passing scores, and operational requirements for test administration, security, and confidentiality as well as requirements for quality improvement	Desk	12 nursing boards; 49 certification programs as of 2006
Conference for Food Protection Standards for Accreditation of Food Protection Manager Certification Programs (Conference for Food Protection, 2004)	• Purpose is to identify the essential components of a certification program to protect candidates and the public • Standards were developed by the CFP; ANSI is the accrediting organization for this standard, designed primarily for certifying organizations attesting to the competency of food protection managers • The standards are based on nationally recognized principles used by organizations providing certification	Desk audit and onsite audit	4 companies accredited as of 2006
Essentials for Approval of Examination Boards in Medical Specialties (American Board of Medical Specialties, 2005)	• Purpose is to maintain and improve the quality of care in the U.S. by assisting specialty boards in their efforts to develop and utilize professional standards for the evaluation and certification of physician specialists • "Essentials" focus on what constitutes a specialty (scope of practice) and the standardization of specialty board practices and procedures; no specific requirements for the technical and administrative aspects of testing	Desk	24 physician boards have been approved as of 2006
Guidelines for Accreditation (Association of Real Estate License Law Officials, 2004)	• Purpose is to assist real estate licensing boards in selecting uniform licensing examinations offered by testing vendors and to raise the quality of national uniform licensing examinations	Desk audit	4 testing vendors as of October 2006

Standard	Description	Audit Type	Accreditation
	• Standards include technical requirements (e.g., passing score and test development procedures, statistical processes) and content requirements (e.g., clear definition of knowledge and skills necessary for practice and satisfactory coverage of knowledge topics at appropriate cognitive levels)		
Guidelines for Engineering and Related Specialty Certification Programs (Council of Engineering and Scientific Specialty Boards, 2000)	• Purpose is to provide criteria for the establishment of specialty certification programs and to provide a quality benchmark for organizations that certify graduate engineers (beyond state requirements). Covers technical and organizational aspects, and includes specific guidelines for how to set eligibility requirements • ESB includes nonengineering organizations that certify engineers and scientific, technical and construction personnel	Desk audit	13 accredited organizations as of 2006
International Standard ISO/IEC 17024 Conformity Assessment—General Requirements for Bodies Operating Certification of Persons (International Organization for Standardization, 2003)	• Purpose is to drive continual improvement and confer additional credibility to certifying organizations, promote their recognition in national and international arenas through a fair and impartial process conducted by a globally recognized third party. Includes technical, administrative and performance management requirements that apply to both the certification agency and any of its vendors • International standard administered in the United States by the American National Standards Institute	Internal audit, desk audit, and onsite audit	As of October 2006, 10 organizations and 26 programs accredited; 27 applications in progress
National Commission for Certifying Agencies Standards for the Accreditation of Certification Programs (National Commission for Certifying Agencies, 2004)	• Purpose is to provide the public a means to identify certification programs that meet competency assurance standards • Includes standards on purpose, governance, and resources; responsibilities to stakeholders; assessment instruments, recertification, and maintaining accreditation • Any type of national certification agency can apply for approval (e.g., healthcare, scientific, business, and finance); approval is by program, so a certifying agency may have multiple programs accredited	Desk audit	62 organizations and 137 programs are accredited as of 2006
Standards for BIACO Accreditation of Proprietary Testing Programs (Buros Institute for Assessment Consultation and Outreach, 2003)	• Purpose is to improve the science and practice of testing for proprietary tests and testing programs • Standards based on the Standards for Educational and Psychological Testing (1999) and Association of Test Publishers Guidelines for Computer-Based Testing (2004)	Desk audit and onsite audit	3 companies accredited as of 2006

TABLE 4.3　Business Standards That Apply to Testing Organizations

Standard	Description	Audits	Programs Recognized
ANSI/ISO/ASQ Q9001-2000 American National Standard: Quality Management Systems—Requirements (International Organization for Standardization, 2003)	Generic quality management standard that promotes a process approach to management and continuous improvement. The standards include requirements for a documented quality policy and objectives, a quality manual, documented procedures for control of documentation, documented procedures for product development and quality control, documented records to show conformity to documented policies and procedures, ensuring availability of resources to support the quality policy, monitoring the policy, and taking improvement actions.	Annual internal audits and annual third party audits	http://www.whosregistered.com lists bodies that have been certified to 9001, for those who choose to be listed
CMMI® Capability Maturity Model Integration (Chrissis, Konrad, & Shrum, 2003)	The goal of the CMMI Model is to increase an organization's ability to reliably develop products and services in a repeatable fashion with continual improvement. It includes models in four disciplines: systems engineering, software engineering, integrated product and process development, and supplier sourcing. The model also includes capability levels (0 to 5) that describe the organization's capability relative to a process area and maturity levels that define levels of process improvement and range from 1 (process unpredictable, poorly controlled, and reactive) to 5 (focus on continuous process improvement). The Software Engineering Institute (SEI) trains appraisers and provides a list of trained appraisers.	Self-assessment or SEI-trained appraiser can conduct appraisal	A voluntary list of appraisal results is published at http://seir.sei.cmu.edu/pars/pars_list_iframe.asp

Standard	Description	Audit	URL
ISO/IEC 27011: 2005 International Standard: Information Technology – Security Techniques—Information Security Management Systems—Requirements (International Organization for Standardization, 2005)	An international standard for "establishing, implementing, operating, monitoring, reviewing, maintaining and improving an Information Security Management System (ISMS)" (2005, p. v). The standard adopts the "Plan-Do-Check-Act" (PDCA) process improvement model and is aligned to ISO 9001 to allow organizations to more easily use both standards simultaneously.	Internal audits; annual onsite audits	http://www.17799central.com lists certifications to BS 7799 and ISO 27001 for those who wish to be listed
Suitable Trust Services Criteria and Illustrations for Security, Availability, Processing Integrity, Privacy, and Confidentiality (Including WebTrust™ and SysTrust™) (American Institute for Certified Public Accountants and Canadian Institute of Chartered Accountants, 2003)	The purpose of this standard is to provide basis for CPA to independently evaluate whether an information technology system is reliable. The reliability is measured against four principles: availability, security, processing integrity, and confidentiality. The standard provides specific criteria that should be achieved to meet each principle. The CPA tests controls; an unqualified attestation report is issued if controls were operating effectively and meets one or more of the rigorous SysTrust standards in the areas of security, processing integrity, availability, confidentiality, or system reliability.	Annual onsite audits	http://www.webtrust.org/abstseals.htm is a voluntary list of organizations that have received or currently have a seal

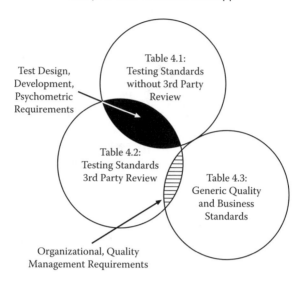

Figure 4.1 Test creation process.

individual psychological associations in the federation do conduct test evaluations (see chapter 8 by Marten Roorda).

All the testing vendors we interviewed, whatever their primary market, said that they are guided in their technical practices by the *Standard for Educational and Psychological Testing.* In fact, many companies state in their promotional materials that they follow and adhere to the *Standards for Educational and Psychological Testing* (1999).

The other standards described in Table 4.1 are more specific in their focus. The Code of Fair Testing Practices in Education (Joint Committee on Testing Practices, 2004) is intended to protect test takers in the educational environment. The International Test Commission (ITC) has taken a slightly different approach to standard setting and posits that appropriate test use will be more likely if test users (as opposed to a focus on test makers) have appropriate knowledge and skills in test taking. The ITC Guidelines on Test Use (ITC, 2001) carries this further by presenting guidelines to help the user identify the knowledge and skills necessary for the specific testing situation. ITC Guidelines on Adapting Tests (2005) was developed to respond to the growing international nature of testing and the need for standards that test publishers can use when adapting tests for different languages.

Both the Association of Test Publishers (2004) and the International Test Commission (ITC, 2005) have developed guidelines for computer-based testing. Although security is covered under these two standards, users feel that security guidelines are not as detailed as they should be, given several major incidents of item and test piracy. The Association for Test Publishers has identified test security as an area for future emphasis in its standards.

The IMS Question and Test Interoperability Specifications (IMS Global Learning Consortium, 2006) contain specifications for developing item banking, test content management, and test delivery tools to maximize the ability of the user to switch from one software product to another. Meeting IMS specifications for interoperability is important if a testing company wants to compete internationally or in specific markets.

The standards in Table 4.1 are not accompanied by a requirement or a mechanism for a third-party review such as accreditation. The use of standards as internal guidelines, without any systematic third-party evaluation that assesses the effectiveness of an organization in meeting or complying with the standards constitutes the most common use of standards. The standards in Table 4.1 serve only to guide/inform the development of testing processes and procedures and sometimes are used to help standardize processes and promote quality improvement within an organization through more formal internal evaluation. For example, the ETS has developed and publishes its own internal audit standards, the *ETS Standards for Quality and Fairness* (2000), based primarily on the *Standards for Educational and Psychological Testing*. Each testing program (e.g. Praxis, GRE, and TOEFL) is periodically audited internally, often every three years. A Visiting Committee, appointed by ETS, of psychometricians, educators, and other stakeholders reviews the results of these audits and makes recommendations that are considered by staff and the ETS Board of Trustees.

Standards Related to Testing and Accreditation

Accreditation is a voluntary process by which an agency (third party) grants a time-limited recognition to an institution, organization, business, program, or other entity that it has met predetermined and standardized criteria. The accreditation process includes the development of the standards and their requirements in a highly systematic way, much in the way described in the ISO definition noted previously. Then these standards are applied through a formal review or audit of the testing program in which experts, usually peers, evaluate the performance of the entity seeking accreditation against the agency's requirements.

We found eight organizations that use standards to accredit testing programs, *all* in the certification/licensure field. They are listed and described in Table 4.2. Interestingly, we were unable to identify any such programs in the educational testing, psychological testing, or industrial/organizational sectors of the testing industry.

Some of the accreditation organizations listed in Table 4.2 are specific to certain disciplines such as nursing (ABNS), medicine (ABMS), engineering-related specialties (CESB), food services (CFP), and real estate (ARELLO). Others such as the American National Standards Institute (ANSI) and the National Organization for Competency Assurance (NCCA) accredit programs in a variety of fields. NCCA is the older of the two accreditors. NCCA began certifying organizations in 1977 and lists more accredited programs than any certification accreditation body in the United States. All accreditation agencies list accredited programs and their status on their Web sites for the public (although the Buros Institute for Assessment Consultation and Outreach [BIANCO] only lists those organizations that wish to be listed).

Audit procedures vary among these accreditation agencies. Some make their accreditation decisions based mainly on a *desk audit*, which involves a review of the application and documentation by trained and qualified individuals against the accreditor's standards and requirements. Only two organizations, ANSI and BIACO, require a site visit conducted by qualified auditors/assessors, in addition to the desk audit.

Ongoing accreditation procedures also vary. For example, the ANSI accreditation is for five years. The initial accreditation requires a desk audit and on-site review. After becoming accredited, an institution undergoes a required annual surveillance process of submitting annual reports to note any changes in the program that might affect compliance with ISO 17024. Assessors review the report and any additional documentation in order to recommend whether accreditation should be continued. An additional site visit is required in the third year of the accreditation, and after five years the program must submit a new application for accreditation. NCCA accreditations are also valid for five years; however, no on-site review is required. To maintain accreditation, an annual report form is required, attesting to the status of the certification program and any changes that have occurred during the year.

Despite the number of programs accredited by these agencies, a 2003 survey of certification organizations conducted by Knapp & Associates International, Inc. and ANSI revealed that only a minority of organizations participating in the survey currently holds accreditations. It appears, however, from the data that the interest in accreditation in this sector is very much on the rise because nearly 40% of the respondents said that they intend to seek accreditation in the near future.

Other Standards of Interest to the Testing Industry

Tables 4.1 and 4.2 describe standards that are specific to testing products and processes. Although standards specific to testing are necessary ingredients to the quality of the testing industry, organizations in our survey felt that testing standards are not sufficient. Table 4.3 describes standards that have been cited by testing companies or test sponsors as being relevant and useful to the testing enterprise. Instead of focusing on a specific industry, these standards are written to be used generically and can apply to the processes used in any company (see chapter 1 for a description of process). All four include a third-party review component and an accreditation-like outcome, although each uses a term other than accreditation. (In the standards world, *accreditation* is reserved for judging that certifying bodies have met a standard. When standards are applied to organizations in general, different terminology is used. For example, organizations become certified against the ISO standards and receive an unqualified attestation report from SysTrust™.)

Several of our survey participants felt that the heavy reliance on information technology in the testing industry suggested the need to use standards related to the information technology aspects of the testing industry. Three of the four standards (SysTrust, CMMI®, and ISO/IEC 27011:2005) relate to information technology.

The American Institute of Certified Public Accountants (AICPA) and the Canadian Institute of Chartered Accountants (2003) have developed SysTrust, a set of principles that can be used to evaluate whether an information technology system is reliable. The system can be evaluated in terms of one or more of the areas of security, processing integrity, availability, confidentiality, or systems reliability. The standard was designed to meet the needs of organizations that are concerned about issues of security, reliability, and privacy of business systems. SysTrust is being used to evaluate the technology system associated with the development, delivery, and scoring of a new computer delivered examination.

ISO/IEC 27001:2005 is a new international standard focusing on "establishing, implementing, operating, monitoring, reviewing, maintaining and improving an Information Security Management System (ISMS)" (2005, p. v). Although 27001 is a new standard, it is based on an older British Standard (BS 7799-2), which companies, especially those working in Britain, have been using. The standard has been designed to be in alignment with ISO 9001 to allow organizations to use both standards simultaneously. ISO/IEC 27001:2005 and its predecessor are being used by organizations to design security systems for their programs, especially when the programs accept credit card information.

The CCMI Capability Maturity Model (Chrissis, Konrad, & Shrum, 2003) includes models in four areas: systems engineering, software engineering, integrated product and process development, and supplier sourcing. The software engineering model may be the area most often used in testing organizations, because that was the first of the four models to be developed. The intent of the model is to increase an organization's ability to reliably develop products and services in a repeatable way and to continuously improve. Testing organizations that contract with the Department of Defense are required to obtain a specified level of maturity on this model.

The most well known of the four standards is ANSI/ISO/ASQ Q9001-2000 (2000). The standard is recognized internationally, and certification to the standard may be required for testing companies in some countries. It is a quality management standard. Several testing sponsors have been certified against this standard.

Several of our survey participants also felt that a focus on continuous improvement is important. Three of the four standards (ANSI/ISO/ASQ Q9001-2000, ISO/IEC 27001:2005, and CCMI) include continuous process improvement models. Continuous improvement models require that an organization is continuously evaluating its processes and improving them from year to year. In chapter 2, Dr. Ramaswamy provides an overview of quality management and quality improvement, and chapters 13 through 17 provide examples of business tools used for continuous improvement.

In total, Tables 4.1, 4.2, and 4.3 represent an armamentarium that testing organizations can use to self-audit their processes to ascertain whether their programs and tests are consistent with best practices, or to submit their programs for third-party audits or accreditation. In addition, they provide the test user or the test sponsor with criteria for evaluating potential vendors.

Why Should Testing Organizations Use Business Standards?

Given the plethora of standards, one can easily assume that standards are "good". However, seriously applying standards in a business costs time and money. Is there research that identifies business advantages to adopting standards? The IMS Global Learning Consortium Inc. sponsors the IMS Question and Test Interoperability (QTI) specifications (see Table 4.1) and publishes case studies of those organizations that have implemented the standards. Pearson VUE (2005) reported avoiding $100,000 in costs by importing test questions for one of its clients using the IMS QTI specifications. This avoided the cost of developing a customized data converter to import items—an expense that is usually necessary with proprietary software that doesn't conform to this standard.

Accreditation standards (Table 4.2) apply to certification bodies. There have been no studies to determine whether accredited certification bodies are more successful than their unaccredited counterparts. However, a NOCA survey of its NCCA accredited organizations (2005) provides some insight into the value of accreditation. Survey results revealed several major benefits associated with continued accreditation: public recognition of meeting industry standards; continued internal quality improvement; enhanced reputation of the program; and defense of the program's integrity in the event of legal challenges.

The survey also shows that a high percentage (50%) of responding organizations felt that accreditation was important for maintaining compliance with requirements by state and local governments and other third parties. (For example, the CFP and ARELLO encourage their constituents to recognize certifications only from accredited organizations. The U.S. Department of Defense requires individuals who perform certain information-assurance functions to be certified by certification bodies that have been independently accredited by a third party.)

The testing industry is small, and there are so few testing companies using the business standards in Table 4.3 that it would be impossible to conduct a scientific study of the value of business standards to the testing industry. Such studies are difficult to conduct even in a broader context. Because the business standards in Table 4.3 are generic, the research from other industries can be used to evaluate the standards' general usefulness. Three of the four standards listed in Table 4.3 have been studied.

Because information technology is such an integral part of the testing industry today, perhaps the most relevant study is one of the impact of ISO 9000 certification and CMM certification on the software industry in India (Issac, Rajendran, & Anantharaman, 2004). The authors developed an extensive survey of quality management practices. This survey was given to employees in software development firms. Responses were compared for noncertified firms, ISO 9000 firms, and CMM firms. The authors found that

> quality certification helps software firms to attain higher operational performance in the form of better product attributes and higher returns on quality. Further CMM (levels 4 and 5) certification of the firms appears to provide a competitive edge over ISO 9000 certification, since CMM-certified firms have been found to have better management practices and higher operational performance. (pp. 23–24)

Charles Corbett, Maria Montes, David Kirsh, and Maria Alvarez-Gill (2002) analyzed financial performance of ISO 9000-certified firms versus a control group of noncertified firms. They compared the results over a ten year period for three business sectors—chemicals and allied products, industrial and commercial machinery and computer equipment, and electronic and other electrical equipment and components. Results show that "firms experience significantly better performance after deciding to seek their first ISO 9000 certification, than a control group of firms with similar performance prior to that decision" (p. 39).

Although the previously mentioned research and our interviews suggest that active use of standards and third-party accreditation can add value to an organization, there are related costs. Each test provider must make an informed decision about how standards will best be used to meet organizational goals.

Conclusions and Lessons Learned

Organizations within the testing industry have most often focused on standards specific to testing (see Table 4.1). In certification testing, third-party accreditation (see Table 4.2) has been used as a way of protecting the public and assuring the quality of a credential. In some cases, certifications are not recognized by the field unless the certifying organization is accredited. The use of other, more generic, standards in the testing industry is rare.

The absence of accreditation for fields such as educational testing, admissions testing, psychological testing, industrial/organizational testing, and test delivery is striking. Toch (2006) estimates that by the spring of 2006, approximately 45 million standardized tests will be administered in grades 3 through 12 to meet the requirements of No Child Left Behind. This does not include other tests given in elementary and secondary classrooms, school admissions tests, or employment tests. Is the field of educational testing so much more advanced than the field of certification testing that it would not benefit from accreditation? This hypothesis seems unlikely given the testing turmoil and testing debates reported in various recent newspaper accounts. Our interviews, however, revealed that these very public exposés have resulted in a shift in attitude on the part of some testing vendors who are serving the K through 12 sector. Public outcry and competition for No Child Left Behind dollars have highlighted the use of standards and audits as a competitive advantage and proof to test users that there is a corporate commitment to quality.

The lack of greater attention toward business standards and standards that emphasize continuous improvement in the educational testing marketplace may lead to government regulation. In May 2006, Senator LaValle of New York filed legislation for the creation of a state board to oversee standardized testing (Arenson, 2006). Toch (2006) recommends an independent national testing oversight agency, to independently audit state testing programs and the testing industry.

Our research for this chapter suggests that the next decade will see more standards and third-party audit programs and that these will be subscribed to by the entire spectrum of the testing community. The competition for revenue, the insistence of consumers, state and federal governments, and sponsoring agencies for quality and the ever-present scrutiny and voice of the media will be the drivers of this change.

It is highly likely that accreditation and certification of testing processes will be conducted against business and psychometric standards using third-party audits. This will be an important tool for customers of the testing industry to differentiate among the various vendors in terms of a commitment to standards and test quality.

Acknowledgments

We would like to thank those who participated in our search for information—your help allowed us to identify major standards, understand their use in the testing industry, and identify issues in the application of standards.

Note

We have attempted to include the most widely used, nonproprietary standards in the United States, knowing that there are probably others that apply to testing products and processes.

References

American Board of Medical Specialties. (2005). *Essentials for approval of examining boards in medical specialties.* Retrieved July 7, 2006, from http://www.abms.org/newbrds.asp

American Board of Nursing Specialties. (2004). *American board of nursing specialties accreditation standards.* Retrieved September 20, 2006, from http://www.nursingcertification.org/pdf/standards_complete_revised_10_04.doc

American Educational Research Association, the American Psychological Association, & the National Council on Measurement in Education. (1999). *Standards for educational and psychological testing.* Washington, DC: American Educational Research Association.

American Institute of Certified Public Accountants, Inc., & Canadian Institute of Chartered Accountants. (2003). *Suitable trust services criteria and illustrations for security, availability, processing integrity, online privacy, and confidentiality (Including WebTrust® and SysTrust®).* New York: American Institute of Certified Public Accountants.

American National Standards Institute, & Knapp & Associates International, Inc. (2003). *Personnel certification: An industry scan.* Washington, DC: Authors.

Arenson, K. W. (2006). Senator proposes creating board to oversee college admissions tests. (2006, May 20). *New York Times.* Retrieved May 25, 2006, from http://www.nytimes.com/2006/05/20/nyregion/20sat. html?r=1&oref =slogin&pagewanted

Association for Assessment in Counseling and Education. (2003). *Responsibilities of users of standardized tests.* Retrieved September 20, 2006, from http://aac.ncat.edu/resources.html

Association of Real Estate License Law Officials. (2004). *Guidelines for accreditation.* Unpublished Document. Montgomery, Alabama: Author.

Association of Test Publishers. (2004). *Guidelines for computer-based testing.* York, PA: Author. Available at www.testpublishers.org/documents. htm

Assurance Services Executive Committee of the American Institute of Certified Public Accountants. (2003). *Suitable trust services criteria and illustrations for security, availability, processing integrity, online privacy, and confidentiality (including WebTrust and SysTrust).* New York: American Institute of Certified Public Accountants, Inc. & Canadian Institute of Chartered Accountants.

Buros Institute for Assessment Consultation and Outreach. (2003). *Standards for BIACO accreditation of proprietary testing programs.* Retrieved September 20, 2006, from http://www.unl.edu/buros/biaco/pdf/ standards 03.pdf

Chrissis, M. B., Konrad, M., & Shrum, S. (2003). *CMMI: Guidelines for process integration and product improvement.* Boston: Pearson Education, Inc.

Conference for Food Protection. (2004). *Conference for food protection: Standards for accreditation of food protection manager certification programs.* Retrieved September 20, 2006, from http://www.ansi. org/conformity_assessment/personnel_certification/apply.aspx?me nuid = 4

Corbett, C. J., Montes, M. J., Kirsh, D. A., & Alvarez-Gil, M. J. (2002). Does ISO 9000 certification pay? *ISO Management Systems,* (July–August), 31–40.

Council of Engineering and Scientific Specialty Boards. (2000). *Guidelines for engineering and related specialty certification programs.* Retrieved September 20, 2006, from www.cesb.org

Council on Licensure, Enforcement, and Regulation. (2004). *Development, administration, scoring and reporting of credentialing examinations:*

Recommendations for board members. Lexington, KY: Author. Retrieved September 20, 2006, from www.clearhq.org/publications. htm

Educational Testing Service. (2000). *ETS standards for quality and fairness.* Princeton, NJ: Author.

European Federation of Psychologists' Associations. (2005). *EFPA review model for the description and evaluation of psychological tests: Test review form and notes for reviewers, version 3.41.* Retrieved September 20, 2006, from http://www.efpa.be

Hambleton, R. K. (2005). Issues, designs, and technical guidelines for adapting tests in multiple languages. In R. K. Hambleton, P. F. Merenda, & C. D. Spielberger (Eds.), *Adapting educational and psychological tests for cross-cultural assessment* (pp. 3–38). Mahwah, NJ: Lawrence Erlbaum Associates.

IMS Global Learning Consortium, Inc. (2006). *IMS question and test interoperability overview: Version 2.1 public draft specification.* Retrieved April 14, 2006, from http://www.imsglobal.org/question/qti_v2plpd/imsqti_oviewv2plpd.html

International Organization for Standardization. (2000). *ANSI/ISO/ASQ Q9001-2000 American national standard: Quality management systems—requirements.* Milwaukee, WI: American Society for Quality.

International Organization for Standardization. (2003). *International standard ISO/IEC 17024 conformity assessment—general requirements for bodies operating certification of persons.* Geneva, Switzerland: Author.

International Organization for Standardization. (2004). *Guide 2: Standardization and related activities—general vocabulary* (8th ed.). Geneva, Switzerland: Author.

International Organization for Standardization. (2005). *International standard ISO/IEC 27001 Information technology—security techniques—information security management systems—requirements (2005-10-15).* Geneva, Switzerland: Author.

International Test Commission. (2001). International guidelines for test use. *International Journal of Testing, 1*, 93–114.

International Test Commission. (2005). *International guidelines on computer-based and Internet delivered testing.* Retrieved January 2, 2006, from http://www.intestcom.org/guidelines

Isaac, G., Rajendran, C., & Anantharaman, R. N. (2004). Significance of quality certification: The case of the software industry in India. *Quality Management Journal, 11*(1), 8–32.

Joint Committee on Testing Practices. (2000). *Rights & responsibilities of test takers: Guidelines and expectations.* Washington, DC: Joint Committee on Testing Practices, American Psychological Association.

Retrieved September 20, 2006, from http://www.apa.org/science/jctpweb.html

Joint Committee on Testing Practices. (2004). *Code of Fair Testing Practices in Education.* Washington, DC: Joint Committee on Testing Practices, American Psychological Association. Retrieved September 20, 2006, from www.apa.org/science/jctpweb.html

National Commission for Certifying Agencies. (2003). *National Commission for Certifying Agencies Standards for the Accreditation of Certification Programs.* Washington, DC: Author.

National Organization for Competency Assurance. (2005, Summer/Fall). *The value of NCCA accreditation Survey.* Unpublished manuscript. Excerpts available in *NOCA News,* Washington, DC: Author.

Pearson VUE. (2005). *Pearson VUE: Exchanging testing materials with partners using the IMS question and test interoperability specification.* Retrieved April 19, 2006, from http://support.org/question/qtiPearsonCaseStudy.pdf

Society for Industrial and Organizational Psychology. (2003). *Principles for the validation and use of personnel selection procedures.* Retrieved September 20, 2006, from www.siop.org

Toch, T. (2006). *Margins of error: The education testing industry in the no child left behind era.* Retrieved February 11, 2006, from www.educationsector.org/usr_doc/Margins_of_Error.pdf

5
The Value of Accreditation

Roger L. Brauer

Executive Director, Board of Certified Safety Professionals

Introduction

Certification has grown rapidly in the United States during the last two or three decades. It provides an option for domains of practice that do not fall under government licensure. It is a means for a professional area of practice to assess and benchmark competency of practitioners.

Most members of a practice area and those relying on a practice assume that a government agency handling licensing of the practice uses reliable processes. For a peer-operated certification, those using or relying on the practice area are not always convinced that a certification has the same reliability as licensing by a government agency. As a result, third-party, independent accreditation of certification programs has emerged to assure stakeholders of the quality of the certification.

This chapter explores the value that an accreditation process adds to certification. The chapter identifies the primary stakeholders, identifies how quality can be identified and accreditation adds value for stakeholders, and discusses the need to help certificants understand quality and be good shoppers for certification. The chapter compares three accreditation programs and the impact of accreditation for a certification body. Throughout the chapter, there are examples of how accreditation has added value for the Board of Certified Safety Professionals (BCSP) and its Certified Safety Professional® (CSP) program. There is also a summary and lessons learned.

The Certification Stakeholders

One needs to understand the stakeholders associated with a certification program. Clearly, the certification body is a stakeholder along with the individuals achieving the certification offered. Other, important stakeholders are employers and users of the individuals achieving the certification. Stakeholders are the membership or professional organizations associated with the area of practice covered by the certification. Government agencies at local, state, and federal levels may be stakeholders. Finally, the public being directly or indirectly impacted by the certified practice is an important stakeholder.

The Certification Body

The certification body is the primary stakeholder. It becomes the *keeper of the list* that defines who has met the established standards for the practice being certified. Its existence, credibility, financial success, and image all depend on how well it handles its role as the primary stakeholder.

The Certificants

The applicants who seek the certification, the candidates who become accepted into the certification process, and the certificants who ultimately achieve the certification are also important stakeholders. Their personal credibility depends on the quality of the certification and the value that it brings to the practice being certified. They pay the fees to pursue and retain the certification and want the use of the credential to be meaningful for others. They want to be in a list that enhances their own image and adds value. When those achieving the certification can say "the process made me learn the subject matter," there is assurance that the certification has added value for those achieving it.

Employers and Other Users of Certified Practitioners

Those who use people in the practice being certified are important stakeholders. These stakeholders may be companies and organizations hiring someone for a position or an organization wishing to contract for services. When a certification requires education or training and experience as a qualification for the certification and thoroughly evaluates such applicant qualifications, the employers have an independent means for verifying these important characteristics of potential employees. When a quality examination process assesses the knowledge required for current practice and when there is a recertification requirement for staying current, employers and those seeking services from practitioners have assurance that the individuals are qualified to perform the functions of the field being certified.

Membership and Professional Societies

Membership organizations and professional societies affiliated with the field of practice covered by the certification are also stakeholders. They may use the certification to differentiate levels of competence. Sometimes an organization allows only those who achieve certification to become members. Other organizations may use certification to allow members to qualify for the highest level of membership status or serve as a director or officer.

Government Agencies

Government agencies may rely on the certification through citations in laws, regulations, standards, policy, or contracts. As a stakeholder, they rely on the certification to ensure that those performing certain functions are qualified for such roles. The reliance may intend to make good use of taxpayer money. In other cases, the reliance may involve protection of the public from harm. Unless a certification is able to show that it has met independent, third-party accreditation, a government entity is not likely to name one specific certification over another in its documents. The accreditation process provides governments with a means to qualify a certification for citation in public policy.

The Public

For most certifications, the ultimate stakeholder is the public. If a practice provides services directly to the public, it is likely that governments will use licensing to protect the public from practitioners. Certification that meets independent accreditation standards also provides assurance to the public that practitioners are qualified to provide services. When practitioners work for employers and the employers are responsible to the public for their products and services, certification conveys competency and quality on behalf of the employer, and accreditation of the certification adds to that image.

Beating the Competition

The level of competition among certifications for one field of practice will vary significantly from that for another field of practice. Accreditation helps elevate the quality of a certification program so that it may differentiate itself from competing certifications.

With the introduction of certification for a field of practice, there is always a counterculture. Generally, the leadership for the field will see a need for a certification program to help ensure competency among practitioners. At the start, a number of people in practice will share that value and pursue the certification as soon as it is available.

At the same time, there will be many in practice who feel that they are good at their chosen practice and do not need any assessment of their personal qualifications. They may feel that they worked hard to learn the practice now covered by certification and do not perceive that certification will add value for them. They may feel that the certification program is simply a means for a new organization to make money from those in practice. Others may feel that the standards for certification are too high. They may not want to risk failing those standards or to work on improving themselves to be able to achieve those standards. In some fields, some may seek to start another certification process that looks similar, has lower standards, and is easy to obtain.

One means for a certification to publicly demonstrate the quality it contains is through third-party, independent accreditation. The quality certification can demonstrate compliance with standards established for certifications by achieving accreditation.

Helping Potential Certificants Shop for Certification Quality

Although third-party, independent accreditation is important, it is certainly not sufficient to be able to compete. The marketplace for a certification must understand the value that accreditation adds.

Those in practice must learn what to look for to understand the difference between an accredited certification and one that is not accredited. They need to recognize that pursuing or planning to pursue accreditation has value. The burden of proof is on the shoulders of the accredited certification program, not on the nonaccredited one, to convince buyers of the quality that accreditation adds. The means for providing this proof is through education of the shoppers, those who are candidates for certification and the other stakeholders who rely on certification and use the services of those who practice in the field being certified.

One approach is to clearly mark the literature for the quality certification with accreditation marks, statements about accreditation, and explanations of what accreditation means. In addition, it may be necessary to train potential buyers to understand the features of accreditation and what accreditation covers. It may be necessary to explain what to look for in a certification program to detect whether it can achieve accreditation.

For example, some certifications offer training courses and award certification to those who complete the training. A good shopper for certification needs to understand that a very important accreditation standard is independence between training for a certification examination and the process of achieving the certification. In other words, accreditation standards typically require that a certification program be independent of preparation for the examination required in the certification.

Another example that provides insight into quality for certifications is the passing score information. Some certifications publish their passing score for an examination as 70%. A flag should go up for a good shopper, because the likelihood is very low for selecting 70% for a passing score through an acceptable procedure for setting a passing score. It is much more likely to be some unusual value that is different from passing score standards typically used in college courses.

Another example is failure to publish the financial information for the certification body. Accrediting standards require that users and stakeholders have access to such information.

A further example is failure to publish who serves in the governance of the organization offering certification and making decisions about who gets certification and who does not. A certification program that offers no information about the officers and directors should be suspect.

Thus, there are many ways one can quickly detect whether a certification program has or could achieve accreditation. It is important to educate the buyers so that they can be good shoppers and be able to identify quality.

Most people may not even know that accreditation exists for certifications or know how to find out if a certification holds accreditation. They may not understand what accreditation involves and what accreditation standards cover. As a result, the buyers and users of certifications need to be educated about organizations offering accreditation.

BCSP began to tackle the issue of accreditation awareness by finding opportunities to explain accreditation and what it adds to the certification. It prepared articles for trade publications and professional magazines on accreditation and what it means. The approach focused on how to be a good shopper for certification. The opportunities to explain accreditation focused on specific features that one should look for in order to determine whether various available certifications were even able to achieve accreditation. After several years and contributing materials for numerous publications, measurable progress was evident when other articles that discussed or commented on certification pointed readers to accreditation as a seal of excellence. Progress became evident when accreditation began to be cited in government regulations and even legislation.

Increasing Value for Certificants

Accreditation provides added value for certificants. It assures them that the certification they achieved or will pursue has met a set of standards for certification programs in much the same way that they must meet a set of standards established for their area of practice in order to demonstrate competency and achieve certification.

Accreditation provides assurance that the certification functions effectively, is fair and consistent for all, is soundly managed, and will continue to exist because of sound management and financial health.

The value may be assessed in various ways. One way is through feedback from those completing the certification process. Quality certifications will be those achieving the following feedback from certificants: [the certification] "made me learn the subject" [matter of practice]. A poor quality certification is not likely to generate such

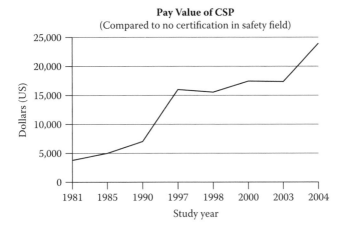

Figure 5.1 Pay value of CSP.

responses from the population achieving certification, because it does not require much to meet a standard.

Another way to demonstrate the value of a certification is through pay differentials between those who are certified and those who are not. Early on, it is difficult to identify the pay differential for those achieving certification. The longer a quality certification is in place, however, the more likely this pay differential will increase.

Figure 5.1 is a plot of the pay differential for the Certified Safety Professional (CSP) that has evolved over 30 years. The data come from salary surveys completed by professional membership organizations, the Board of Certified Safety Professionals (BCSP) itself and by publishers of trade magazines in the field. Over time, the pay differential of the certification has grown significantly.

A value that a quality certification provides in the United States that is often overlooked is lateral mobility. Lateral mobility refers to the ability of someone to find a position in a field of practice with another employer when leaving one employer for any of a variety of reasons. In general, certification in the United States has aided lateral mobility.

Compare the typical employment model of a generation ago with that which is more typical today. A generation ago, someone entering a professional field not covered by licensure usually sought a position with a major employer and stayed with that employer until retirement. Most people who entered a career field stayed in that field their entire career.

With downsizing, rightsizing, and capsizing of employers and sales of major companies, that model is not likely to apply today. In addition, the rate of change in most fields of practice today often leads people to seek other kinds of practice and change career direction. Some have predicted that the typical college graduate today will change careers five or six times during their employment lifetime.

Certification allows people to demonstrate competency in a field in which they work or seek to work. When forced to find alternate employment or when choosing to do so, certification provides a means to demonstrate competency in practice. Employers who rely on a quality certification, one holding accreditation and recognized by stakeholders who use, contract for, or employ practitioners in the certification's field, will consider certification as an important attribute in selecting someone for a position. Education, experience, demonstrated leadership, strong references, and other factors are also important.

The Board of Certified Safety Professionals has tracked job advertisements in a professional journal in its field for more than 20 years. The data show that requirements for education and experience typically appear in about 90% of these ads. The rate for these two characteristics has been fairly constant. During the last 30 years, however, the ads that require or prefer the Certified Safety Professional credential has increased from about 20% in 1980 to over 50% in the last 5 years.

Selling Professionalism versus Selling Certification

Associated with a focus on quality processes and achieving accreditation is another lesson that BCSP has learned. It is learned that selling certification often is not enough. The role that certification and the quality that accreditation adds is that of selling professionalism, not just the certification. In short, many people are not simply interested in the certification alone, but understand that certification can play a major role in advancing their own professionalism associated with the practice covered by the certification.

In many fields, education or training and experience define the entry point. Most employers include education or training and experience qualifications in their ads for positions. Certification may become an additional qualification once it becomes established. When competing for jobs or contracts, holding certification in an area of practice may provide a competitive edge. Certification often provides a competitive edge when someone is considered for advancement or a higher level of responsibility and pay.

As a result, certification adds to the path of professionalism and career growth. An accredited certification increases value in one's career path and helps to define professional progress. Holding an accredited certification helps define professionalism in the field for which the certification is awarded.

Increasing Value for Other Stakeholders

Accreditation of a certification helps establish the quality that the certification represents. It adds value for all stakeholders. It provides assurance that it is soundly managed and does what a certification program should do. It enhances confidence in the certification, the competency of those achieving certification and the overall image of the people who achieve the certification.

A quality certification should be able to provide value to other stakeholders. Some of these values may be difficult to measure directly or precisely.

Quality certifications typically establish qualifications for candidates. Most often there are educational or training and experience requirements to be eligible to sit for the certification examination(s) or to achieve the certification. These qualifications and their consistent enforcement provide value to stakeholders. The certification body should clearly explain these requirements and how they are evaluated. Employers and contract issuing agencies can rely on the certification process to ensure compliance with education/training and experience standards.

Consider the Board of Certified Safety Professionals as an example. Its policies and application review procedures help assure stakeholders that degrees presented on applications are legitimate degrees and were awarded to the applicants. BCSP requires that candidates for the Certified Safety Professional have at least a bachelor's degree in any field or an associate's degree in safety and health. BCSP applies a sliding scale to determine the amount of credit awarded in the application process for different degree fields. A bachelor's degree in safety holding program accreditation from the Applied Science Accreditation Commission of ABET, Inc., receives maximum credit. Bachelor's degrees in other fields receive less credit. The BCSP staff verifies each degree through an official transcript from the issuing school; copies are not allowed. The transcript must show that the degree was awarded and the date. The accreditation status for a safety degree is checked against the official list published by ABET, Inc. More detailed requirements exist for associate degrees in safety and health and the staff checks courses on transcripts against course standards for these degrees.

In addition, BCSP established standards for degrees to ensure that they are not from diploma mills. To be acceptable to BCSP, a degree from a U.S. college or university must have been awarded during a period during which the school held institutional accreditation from a body recognized by the U.S. Department of Education or the Council for Higher Education Accreditation. To verify the institutional accreditation of a school, BCSP staff refers to the listings on the Web sites of the two organizations. If a transcript is from a school that is not found on these Web sites, the staff refers to other sources. One source is the Web site operated by the State of Oregon, Student Assistance Commission, Office of Degree Authorization, and its list of unaccredited colleges and other publications on diploma mills. On occasion, there is a need to research a school that has closed, but previously held institutional accreditation.

Applicants must have four or more years of *professional* safety experience. BCSP uses five criteria to evaluate professional-level experience in safety. Each period of experience is evaluated separately. Failure to meet any one of the five criteria results in no credit for the period of experience.

On BCSP applications, candidates must disclose any previous criminal convictions. BCSP has a policy related to three categories of criminal convictions. The lowest, minor convictions do not prevent the application from being accepted. The middle level involves a range of convictions that require review by at least four people who must agree on the case to determine if the application will face a delay or not be accepted. The third level involves any felony, which results in denial of the application. In a number of cases, BCSP relies on its attorney to verify the actual court records so that no error is made in the decision on acceptance of the application.

Through these clearly defined and published standards and procedures, employers and other stakeholders have assurance that the certification process covers candidate attributes other than knowledge by examination. The stakeholders have assurance that those who achieve the Certified Safety Professional certification possess a range of minimum qualifications for professional practice.

Membership Organizations

Accreditation helps enhance the profession or practice covered by a membership organization. Accreditation helps the membership organization financially by providing opportunities for the preparation of individuals for certification. It provides opportunities for providing continuing professional development after certification. Accreditation eliminates the conflict of interest resulting from a certification body handling both the preparation for certification and the certification itself. A certification body that provides preparation for a certification and awards certification upon completion of the training would not be able to meet accreditation standards.

Employers and Other Users of Certificants

Accreditation of a certification allows employers and others to be more confident in using the certification as a selection factor for potential employees, contracted work, or promotions.

Government Organizations

Accreditation provides an independent evaluation process that frees a government organization from having to establish its own criteria for determining whether it can cite a certification. A dilemma for a government organization that wants to assure competency among practitioners cited in laws, regulations, and standards or in contract solicitations is being able to defend the citation of a certification for that field of practice. Simply naming a certification appears arbitrary and can lead to challenges from other certifications. Relying on an independent, third-party accreditation process for certifications sets clearly defined qualifications that the government organization can use when seeking to use certification to protect the public or to assure that minimum competency has been met.

Accreditations and What They Cover

Organizations with Accreditation Standards for Certifications

In chapter 4, a number of accreditation standards for certifications were presented. BCSP chose to become accredited by three standards that are particularly relevant for the CSP. The three accreditation bodies include the National Commission of Certifying Agencies (NCCA), the Council of Engineering and Scientific Specialty Boards (CESB), and the American National Standards Institute (ANSI), which administers the ISO/IEC 17024 program covering certification of persons.

NCCA began in the 1970s with a grant from the Department of Health, Education, and Welfare. Initially, its standards covered allied health fields, but it soon emerged as an organization that accredits nearly any certification at any level of practice. It operates independently under the National Organization for Competency Assurance (NOCA), an organization with certification bodies and vendors servicing the certification industry as its main members. In 2003, NCCA completed a major overhaul of its standards following a comprehensive, two-year study.

In the 1980s, CESB resulted from a national symposium to address the development of competency assessment programs for engineering and related fields outside of state licensing of engineering practice. It adopted many of the early NCCA standards with some variations to them. As indicated by its name, CESB's domain covers engineering, engineering-related, and scientific fields.

In 2003, the International Organization for Standardization (ISO) and the International Electrotechnical Commission (IEC) established ISO/IEC 17024, *General Requirements for Bodies Operating Certification Systems of Persons*. To date at least 85 countries have adopted this standard. The American National Standards Institute (ANSI) administers this standard in the United States. In January 2006, ANSI and ASTM International announced the adoption of the standard as an American National Standard. This standard is applicable to any program handling certification of persons (as opposed to equipment, processes, etc.).

What the Standards Cover

There are considerable content overlap and similarities among the three accreditation standards. All three cover aspects of governance, financial condition and disclosure, fairness to candidates, non-discrimination, independence, participation by stakeholders, examination development, administration and scoring, recertification, and other factors. Certification organizations must file applications for each accreditation, and assigned teams evaluate the information provided with the applications.

There are also significant differences among the accreditation standards in content, depth, and enforcement. Table 5.1 summarizes and compares the contents of the three accreditation standards in general. The comparison is merely a summary and reference should be made directly to the standards for a detailed, comparative analysis.

CESB (2000) standards are quite short. Those of NCCA (2003) and ISO/IEC 17024 (2003) are much more explicit and describe more thoroughly the specific information covered by the standards and the concepts behind them. In addition, the latter two both provide interpretive information to help applicants understand how to comply.

There are also significant differences in contents. Both NCCA and ISO/IEC 17024 cover a wider range of performance elements than do the CESB standards. In general, NCCA has more details on examination matters and psychometrics, such as methods for content validity of examinations, reliability, establishing cut scores (passing scores), examination performance, and other examination matters. NCCA and ISO/IEC 17024 require considerable documentation of job analysis and cut score studies.

The ISO/IEC 17024 standard has strong coverage of management systems, such as thorough documentation of program practices and procedures. Although compliance with ISO 9001 is not required, the ISO/IEC 17024 standard includes many aspects and concepts associated with management systems conformance. ANSI publishes an interpretive document related to examination matters and psychometrics covered by the standard. The certification body applying for accreditation under this standard must thoroughly document

TABLE 5.1 General Comparison of Three Accreditation Standards

Topic	ISO 17024	NCCA	CESB
General			
Nondiscrimination	√	√	√
Discipline (loss of certification)	√	√	√
Notice of changes	√		
Nongovernmental			√
National/international scope			√
Organization/Governance			
Independent/impartial	√	√	√
Authority/purpose of organization	√	√	√
Authority for decisions	√	√	
Financial condition	√	√	√
Insurance	√		
Not doing training for certification	√	√	
Appeal process management	√	√	√
Complaint process management	√		
Appropriate staff	√	√	√
Board election procedures			√
Stakeholder input	√	√	√
Directory of certificants		√	√
Representatives of all certifications		√	
Certification Scheme			
Recognized testing methods	√	√	√
Content validity studies	√	√	√
Content validity study frequency			√
Basis for qualifications	√	√	
Membership not a qualification	√		√
Fairness to candidates	√	√	
Bachelor's degree required			√
Management Systems			
Defined policy	√	√	
Quality & quality objectives	√		
Staff training	√		
Documented management system	√	√	
Internal audit system	√	√	
Security	√	√	√
Contracted Services			
Contracts for outsourced work	√		
Confidentiality for contractors	√		
No conflict of interest	√		
Ensure competence of contractor	√	√	
Qualified staff	√	√	
Records			
Record keeping system	√	√	
Record retention policy	√	√	
Annual report			√

TABLE 5.1 General Comparison of Three Accreditation Standards

Topic	ISO 17024	NCCA	CESB
Confidentiality			
Confidentiality of all certification information	√	√	
Confidentiality of examination results only			√
Staff			
Defined qualifications	√	√	
Compliance with policies	√		
Clear job descriptions	√		
Training and professionalism	√	√	
Examiners (for judged competencies)			
Training and competence	√		
Impartial and objective	√		
Candidate Process			
Complete candidate information	√	√	√
Application process	√		
Fair evaluation of data	√	√	
Evaluation			
Accommodate special needs	√	√	
Verification of qualifications	√	√	
Objective evaluation process	√	√	√
Competence evaluation required	√	√	√
Waiver of exam exception (temp)			√
Test development and reporting	#	√	√
Test sites			√
Test results timely			√
Test performance		√	
Cut score studies	#	√	
Reliability	#	√	
Test form equivalence	#	√	
Decisions			
Decision makers independent	√	√	
Error prevention procedures	√		
Certificate required	√		
Recertification			
Recertification required	√	√	√
Recertification time limit			√
Recertification activities defined			√
Comply with current standards	√		
Use of Certifications/Marks			
Manage use of certification	√		
Protection from misuse	√		
Code of conduct	√		
Protection of certification	√		
Restricts certain certification titles			√
Emeritus/retired status allowed			√

(Continued)

**TABLE 5.1   General Comparison of Three Accreditation
Standards (Continued)**

Topic	ISO 17024	NCCA	CESB
Accreditation Process			
Application for accreditation	√	√	√
Documentation of compliance	√	√	√
Onsite audit	√		
Nonconformance action plan	√		
Annual update report	√	√	√

Notes: √ = standard covers topic in some degree; # = covered during evaluations, but not cited explicitly in the standard itself.

Sources: The sources for the standards are as follows: NCCA, www.NOCA.org/NCCA; CESB, www.CESB.org; ISO/IEC 17024, available for purchase from ANSI, see www.ANSI.org.

policies, administrative procedures, and practices, because the organization applying for accreditation will undergo a review for conformance with its own practices.

In a few subject areas, CESB standards get into details about how to comply, whereas the other accreditation standards require the certifying body to explain and defend its particular practices for similar subject areas.

When applying for accreditation, the documentation required by NCCA and ISO/IEC 17024 far exceeds that required by CESB. The evaluation teams assigned by NCCA and ISO/IEC have much more certification experience, psychometric expertise, and evaluation training than do the teams assigned by CESB. CESB relies on representatives from its accredited certification bodies to handle the evaluations. Their experience in certification matters may be quite limited. NCCA also relies on volunteers, while ANSI provides a basic fee for those serving as members of accreditation teams. The experience in certification and testing matters for a team is an essential consideration when composing the teams. For CESB and NCCA, the accreditation decision is based on the materials included in the application. The ISO/IEC 17024 accreditation team not only reviews submitted materials, but also conducts a site visit and confers with all levels of staff to ensure that compliance with standards is actually practiced, not just claimed by the management level. The site visit also confirms compliance with the organization's written procedures.

For NCCA and ISO/IEC accreditation to be awarded, non-compliance with standards must be resolved prior to award of accreditation, whereas CESB awards accreditation prior to resolution of all nonconformance matters.

All three programs require annual reporting, but the level of detail required in annual reporting varies significantly. Each requires reapplication for accreditation every five years. Only the ISO/IEC 17024 accreditation, however, involves a site visit near the middle of the accreditation cycle.

Improving the Certification Organization through Accreditation

A certification body may set a goal to achieve accreditation. People often focus on the accreditation itself. Very often, the process leading up to the accreditation is the most important part. Going through any external review provides a means to improve. Whether it is a thorough financial audit or an accreditation, the process drives improvement. Preparing for accreditation allows a certification body to identify opportunities to improve and to meet them before the independent, third-party review. In addition, the review itself will often produce recommendations for improvement, even when compliance is not an issue.

Keys to Achieving Accreditation

To be successful in the accreditation process, it is important to have in place good practices and to have written explanations covering how the organization handles compliance with standards. Well written policies and procedures become the institutional memory that carries the organization through changes in personnel and ensures that its own standards are handled consistently and effectively. Developing written policies and procedures is a key to quality.

A good way to get started is to focus on the processes used to move applicants and candidates through to certification and the processes used for certificants after they achieve certification. The processes should include policies and procedures related to denying an application, handling discipline and potential termination or removal of certification and many other contingencies that do not apply to every applicant, candidate, or certificant.

Process Design and Documentation   For the Board of Certified Safety Professionals, documentation began when there was an interruption of key personnel. The knowledge of how various situations were handled walked out the door with the loss of those personnel, and the task for remaining or replacement staff was to figure out how things worked. A goal became that of documenting the processes and creating an institutional memory.

The approach included developing a flow chart that tracked the path of someone from application to loss of certification. The entire process was organized into logical units, which one might call phases or functions. Within each phase or function the focus was on defining the tasks associated with each, the order in which tasks were done, who did them, and what information was needed or produced. The analysis included collecting or defining every document, publication, label, report, form, envelope, or communication that occurred along the way.

For certain phases, the analysis included working with psychometric contractors providing examination development and administration services to establish or explain the practices the organization used. It involved compiling board decisions and rulings for the certification standards, appeals, and other decision points in the processes. It involved documenting critical elements such as examination security, confidentiality of candidate information, and similar matters. It required the collective knowledge and experience of the staff, board members, contractors, and others.

Over time, the written documentation was checked against accreditation standards and other sources of information about certification practices to identify gaps, weak points, and vulnerabilities. The written documentation was expanded to cover solutions to these difficulties, often after working with governance committees and consultants to establish polices and procedures to cover the gaps. Individual cases provided opportunities to reflect on policies and procedures and to establish standard practices that were not already in place.

Along the way, general business practices were incorporated into the documentation in order to establish an organizational culture for excellence, to provide outstanding customer service, to focus on continuous improvement opportunities, to ensure financial success, and to elevate the knowledge and skills of the staff.

A focus on review of processes never ends. It is never completed. It is continuous. Changes in the policies of the certification, improved customer focus, changes in technology and customer expectations, changes in business methods, and many other factors continue to force improvement for how things are done.

Pursuing accreditation became one of the driving forces. Each time there was a need to prepare an accreditation application and to defend how the organization achieved compliance with the accreditation standards; the result was a check of the institutional knowledge and memory of the organization. If it was not clearly written such that existing documentation provides a clear explanation of how the BCSP organization handled its business, there was an opportunity to improve the institutional knowledge and memory.

In summary, it was not the accreditation alone that was important; it was the process of preparing the organization to explain how it worked and to defend itself against external evaluation. The documentation provided evidence of compliance or provided the basis for the evaluation team to offer suggestions for improvement from their own experience, even when there was not an issue of noncompliance. Achieving the accreditation was a benchmark for the certification organization.

Over time, BCSP pursued and met the standards of all of the previously mentioned referenced accreditations. In addition, it underwent annual scrutiny of a board audit, financial audits, and occasional critiques by others. All of these reviews resulted in improving the organization and the certification it offered.

Process Scheduling   Part of the documentation process was helping to define or clarify the scheduling of various phases and tasks of certification. Reviews found that staff may not understand how important certain schedules were to the success of the certification and how schedules might minimize the extra work that resulted from getting cross-wise in various ways with some applicants, candidates, or certificants. Clarifying and charting key dates and scheduling in the certification process were part of the documentation process. Study revealed that by adjusting the implementation dates for certain tasks performed in one department, there was an opportunity to reduce work in other departments. Scheduling requires continuous analysis, just as the rest of the process evaluation does.

A major change in scheduling resulted when BCSP switched from offering paper-and-pencil examinations twice per year to offering computer-based testing. Although it is possible to make that change without adjusting the general testing cycle and windows for taking tests, BCSP made a switch from a batch process to a continuous one. In a batch process, applications had to be in by certain dates prior to testing dates. Then the staff scrambled to get all of them processed in time for the examination dates, to get examination sites arranged and to assign candidates to them. That cycle occurred twice per year.

A number of other processes not related to the application, qualification, and examination portions of the certification remained on annual cycles. Examples are billings for annual renewal fees and recertification reporting.

The switch to a continuous process for candidate flow was difficult, but provided maximum convenience for applicants and candidates. They could apply at any time. They could sit for examinations at any time. They could achieve certification whenever they individually met all requirements for certification. The impact for the BCSP staff was being able to handle any of the tasks in this part of the certification process on any given business day. With the increased opportunities to sit for examinations or re-examinations at any time, the backlog of people in process dropped by about 80% over the next five years. People could move through the two-examination process much quicker in the continuous process.

The continuous process resulted in other policy and procedural changes. For example, in order to ensure that people completed the process and did not get stuck, time limits were added. For the most part, the time limits affected less than 15% of the candidates. Most candidates move effectively through the certification process on their own. The time limits may have helped them manage better. The change to a continuous process allowed some candidates to complete the two-examination certification process in only a few weeks instead of several months.

Such changes also impacted the knowledge and skills of staff and the process documentation. The changes improved the quality of the application, examination, and general certification procedures.

Process Metrics   Another important part of the focus on processes involved establishing metrics that measure what is going on and how well policies and procedures work. This, too, can be a major cultural change. Too often an organization and its staff are delighted that they complete their work. It is important, however, to establish metrics or measures of performance that provide insight into what works and what can be improved. Metrics can help allocate resources to where they are needed and to adapt to sudden peaks or drops in tasks.

Metrics may include simple counts of what is taking place each day, week, month or year. An example is how many applications were received during a time period. The counts, if on a weekly basis, can be a measure of whether the rate of applications is close to the budgeted level or whether there is a significant change in work load.

Metrics can include tracking of average process time. For example, BCSP identified that a change in procedures could reduce the application process time from an average of five weeks to an average of three or four days. A review of the procedures showed that a major change in who has responsibility for application decision making could improve customer satisfaction and a much shorter process time while retaining decision accuracy and consistency. The change also reduced staff work.

Metrics can provide feedback on how well policy is working. BCSP introduced an annual analysis of how certificants achieved their recertification credit through its ten categories of activities. The analysis showed not only how the credit was earned, but identified opportunities to encourage better professional development. The information confirmed that the recertification program was fair, achievable, and met its overall goals.

BCSP also implemented an annual review of how new certificants made it through the certification process and met qualifications. The review started to show the impact that the time limit policy once had. It gave insight into how the shift from a batch to a continuous process improved candidate process time. Every year, there are some who can complete the certification process in less than two weeks, whereas the previous batch process had a minimum of six months. The review identified trends in demographics and provided some feedback and ideas for marketing strategy.

Metrics are an important part of process management and contribute to the ability to demonstrate compliance with accreditation standards.

How Seeking and Maintaining Compliance with the Standards Helps a Certification Body Improve The examples mentioned previously illustrate that pursuing and achieving accreditation is not only valuable for the various stakeholders, but also helps encourage improvements in the certification processes.

Achieving an accreditation also drives continuous improvement. The accreditation programs discussed previously vary in the information required for annual updates, periodic reviews, and application for reaccreditation. One aspect of annual reporting involves identifying what has changed in the certification program. It may require explaining why the changes were made and how they continue to ensure compliance with the accreditation standards.

Conclusions and Lessons Learned

Most will assume that having individuals go through a certification process and achieving the certification adds value for stakeholders. One can make the same assumption for accreditation of certifications. In combination, one would assume that an accredited certification is the most valuable. The proof is, however, in the details.

To claim quality and to add real value for its stakeholders, a certification program must establish clear standards for practitioners and have clearly defined and effective administrative and examination processes that are fair, consistent, open, valid, reliable, and effective in differentiating among those seeking the certification and ensuring that people stay current in the practice for which they become certified. The program and the associated practices and examinations must be defendable against challenges.

The same is true for the accreditation of certifications. An accreditation program must have standards that address quality attributes for a certification and quality processes to evaluate the certification programs seeking the accreditation. The stakeholders for the certification and for accreditation are often the same. The accreditation process must be comprehensive, fair, consistent, thorough, and reliable. The accreditation process must assure stakeholders that any certification achieving accreditation has met the quality standards represented by the accrediting body.

When used together, the combination of certification and accreditation helps assure stakeholders that the process for evaluating competency relevant to practice for the field being certified is a quality process. Together, they have a multiplier effect and add high value.

It is also true that the consequences of failure can destroy value. The failure of the certification process for a single candidate can diminish the value of the certification for all certificants and stakeholders.

The failure of the accreditation process for a single certification can diminish the value of all certifications that it accredits. The main lesson BCSP has learned is that accreditation provides an independent, third-party verification of the quality represented by the CSP certification program. In addition, BCSP has used accreditation to push continuous quality improvement as additional accreditations were pursued. Each accreditation has somewhat different requirements and has required BCSP to focus on additional quality elements in its policies and operations. BCSP also learned that accreditation has value for its stakeholders and increases acceptance and recognition for the CSP.

References

Council of Engineering and Scientific Specialty Boards. (2000). *Guidelines for engineering and related specialty certification programs*. Retrieved September 20, 2006, from www.cesb.org

International Organization for Standardization. (2003). *International standard ISO/IEC 17024 conformity assessment—general requirements for bodies operating certification of persons*. Geneva, Switzerland: Author.

National Organization for Competency Assurance. (2003). *National Commission for Certifying Agencies Standards for the Accreditation of Certification Programs*. Washington, DC: Author.

6

An International Case Study of Accreditation

Peter Kronvall
Certification Division, Swedish Board for Accreditation and Conformity Assessment (SWEDAC)

Introduction

The Swedish Board for Accreditation and Conformity Assessment (SWEDAC) is the national accreditation body of Sweden. As manager of the certification division at SWEDAC, I have been involved with accredited certification of persons since 1994. During the development of ISO/IEC 17024, I was a member of ISO CASCO WG 17 and later chaired the IAF task force (the International Accreditation Forum) which produced the guidance to 17024. The purpose of this chapter is to point out difficulties in introducing certification of persons in different schemes[1] and give some examples on methods to overcome those problems. The chapter is rounded off with some practical advice.

Early History

In order to perform a specific task or service, a person generally must satisfy specific requirements regarding education and/or experience. Sweden is not an exception in this regard. Looking back, it is easy to find examples of *governmental approval*. Some of these examples have similarities to today's system of accredited certification. This chapter describes three case studies and addresses their progress, or lack of it, when moving into accredited certification according to ISO/IEC 17024.

The Swedish construction branch is a good example, as this area historically has been safeguarded by legislation and prescriptive rules. The legislation covered not only how and where to build, but also the competence of, for example, the engineer responsible at a construction site. Similar requirements were put in place for technicians dealing with ventilation systems in new and existing buildings. Both schemes put requirements on specific education, experience, and training. The training must be given by approved training course providers, who also gave the examination and scored the students. The final approval was made by the local authority for building and construction. Completion of this process gave the candidate an approval to work within your local county.

The route to a general approval valid for the whole of Sweden was somewhat different. The National Board of Housing Building and Planning had appointed a specific company to perform the examinations and the approvals. Neither of the two routes described previously were in line with accredited certification. Issues like impartiality, competence, or influence from interested parties were not considered or evaluated in any of these schemes.

Two examples from other sectors are also worth mentioning: the approval of welders and the approval of technicians for nondestructive testing. The basis for these approvals was in broad terms an approval of the company of the welder or the technician.

Problems of the First Accredited Schemes in Sweden

The determining factor for the development of accredited certification in general in Sweden was the decision to join the European Union.[2] From the point of view of conformity assessment, this meant that Sweden had to rearrange the system of National Testing Institutes. This was mandatory. But the Swedish parliament went further than just rearranging the mandatory parts. The Members of Parliament decided to open up a vast number of closed approval systems and turn them into open approval systems, based on the principles of accreditation and conformity assessment. This decision made a great impact on the development of conformity assessment in a number of fields in Sweden. The basis for accreditation was at that time the EN 45000 series of standards, and explicitly the EN 45013 standard for certification bodies for personnel.

SWEDAC started to accredit certification bodies in 1990. The first body to be accredited was Det Norske Veritas for certification of quality management systems, 1991-06-12. The first certification body for persons was accredited, 1991-11-22 for certification of personnel in nondestructive testing. The growth of accredited certification bodies for persons in Sweden is moderate. Today, 15 bodies are accredited for roughly 25 different schemes. The majority of Swedish certification bodies for persons are accredited by SWEDAC, and all except one body were previously accredited according to EN 45013. By the time the transition period for implementing ISO/IEC 17024 ended in April 2005, all of the accredited certification bodies had managed to *transform* their previous work methods according to the demands of the new ISO/IEC 17024 standard. The transition was quite smooth and caused no major problems for the accredited certification bodies.

Returning to the previous examples of welding and nondestructive testing, the development toward accreditation for certification of persons was specifically quick in these two spheres. They already had an established tradition of evaluating and approving personnel. In both cases, European standards of competence requirements had been developed in order to serve as normative documents of the certification process. The starting point toward an accreditation system was on the one hand the EN 45013 and on the other hand the European standard for welders/technicians for nondestructive testing. Would these two fit together?

The accreditation body had the disadvantage of trying to change an established professional routine. Parts of the welding/nondestructive industry's *way of doing things* were transferred to the accreditation system. This included both good and not so good practice.

Welding

A characteristic of the development of the welding scheme was the involvement of the employer of the physical welder. A Swedish welding company was obliged to have a responsible welding engineer. This gentleman turned out to be one of the key persons of the whole process: the initial certification process as well as the surveillance and reassessment processes.

When a welding company had identified a need to certify a welder for a specific purpose, the welding engineer sent the welder to a suitable training course. The welding engineer contacted an accredited certification body and applied for certification in the name of the welder. The parties agreed on a time schedule for the practical examination of the welder. After the welder had completed the training course, the accredited certification body's representative performed the examination on site. The result was evaluated by the certification body, and if the test result matched the requirements, a certificate was issued. The certificate was handed over to the responsible welding engineer at the welding company. Normally the welding engineer stored the certificate in the company safe. The certificate was not in the possession of the welder. In some cases, the welder was not even aware of having passed a certification process, nor of the actual requirements for obtaining and maintaining the certificate. This was all in control of the welding engineer.

The next step was the surveillance of the welder. The system required a competent person to sign a certificate confirming that the welder performed welds according to the requirement of the certificate. Normally the welding engineer signed the certificate every six months. If the certification body didn't hear from the welding engineer, the assumption would be that everything was in accordance with the requirements and that the certificate was still valid. After two years, it would be time for recertification. In order to achieve a renewed certificate, the old certificate had to be signed every six months by a competent person. The application for renewal had to include objective evidence of the welder's performance. This was normally solved with a test report from nondestructive testing. Comparing this process (used in the 1990s) with the requirements today in ISO/IEC 17024, an observer can easily identify a number of gaps. At that time, however, we lacked the tools to do it differently.

Nondestructive Testing

Compared to welding, as described earlier with nondestructive testing, the problems differ somewhat. The basis for certification in nondestructive testing was a European standard as well: EN 473. However, neither the companies/persons nor the authorities of the Nordic countries involved with nondestructive testing found the requirements of the standard applicable for certification. They were used to more specific requirements on training experience and how to perform the examination.

A number of documents in the *NORDTEST-series* were produced in order to support the system with more robust requirements. This set of documents was—and still is—very ambitious. Among many other requirements on how to interpret EN 473, the documents also contain very specific requirements on the examiners.[3] To put it briefly, you must be a certified technician for nondestructive testing on the highest level. Taking into account the very specific requirements on training the candidates and the small size of this branch, Sweden ended up with a system in which hardly any of the examiners were impartial with respect to the candidates. This is perhaps a specific problem for a small country, but it might appear in big countries with many remote locations, for example, Australia and Canada. Later I discuss how we worked with those problems and tried to overcome them in the welding and nondestructive testing branches. But let us first return to the construction branch, where we faced a new set of problems.

The Construction Area: A Different Set of Problems

In 1995 the Swedish legislation for building and construction underwent a major rewrite. The former *responsible engineer* of a construction site was history. The position was replaced by a person responsible for checking that the building was built according to the agreement between the local building/construction authority and the proprietor. One of the rationales for this change was to give the proprietor clear responsibility for the construction process and the final structure. Legislation further demanded that the quality manager/responsible person for quality work be certified by an accredited certification body.

The responsibility to develop the normative documents for certification and the certification scheme was given to the National Board of Housing Building and Planning. The normative document was written in a generic form and put requirements on education, work experience, knowledge of building/construction legislation, knowledge of quality management systems, and general suitability for the task. The requirements for quality management systems, for example, were expressed in terms related to the length of a training course.

Whether an accreditation body, a certification body, a candidate for certification, or a training course provider, the main problem for all involved was that we didn't understand what the requirements were. Neither did we know how the candidates should be examined, the type of examination, the content of examination, or the extent of the examination. Surveillance and recertification were not addressed at all in the normative document.

In order to complete my description of various types of problems we faced in the mid 1990s, I return to the technicians working with ventilation systems. In this case, there was only one change from the old school (where a specific company was appointed to examine and approve technicians for ventilation systems): The examination and approval needed to be performed by an accredited certification body. The normative document, the requirements, and the process were identical. Remember that this normative document was not developed in order to fit in a system with accredited certification bodies. It suffered from the same weakness as the normative document for quality managers. Furthermore, there was another complication. More than 1,000 persons had been approved for the same requirements in the old system. They were now forced to go through yet another process to be requalified.

How Did We Deal with the Problems?

These various types of problems/questions in the fields of construction, welding and nondestructive testing were our reality. We faced a lot of work in a field suffering from a shortage of specific experience. The requirements for certification bodies came from the EN 45011, EN 45012, and EN 45013. One advantage at SWEDAC was that we were active with accreditation in all different fields of certification. We worked with all different types of certification bodies. The applicable standards in the EN 45000 series were structured in the same way, and the content was quite similar. The basis was the same independent type of certification. Another strength SWEDAC had and still has is our extensive international involvement. The relevant organization in Europe is the European Cooperation for Accreditation (EA) and in a global perspective it is the International Accreditation Forum (IAF). In other words, we were quite confident with accreditation of certification bodies in general.

How did we tackle the problems? Well, it was not an easy fix; this was a process that continued for many years. We took it step by step, in order to improve the general system of certification of persons as well as the specific certification schemes. One important factor—which is more or less typical for Swedish certification bodies for persons—is that they are quite small organizations with a background from the technical field of the relevant certification. Consequently, they have limited resources for the development of the certification body, limited knowledge of applying quality management systems, and a strong focus on technical issues.

As I previously mentioned, one of the problems we faced in the welding area was the six months signing of the certificate. This process was completely in line with the requirements of the normative document for certification of welders. The certification bodies did not see the process as a problem; it worked more or less as it had before. The only visual difference was the certification and recertification done by them. The control of the fulfillment of the certification criteria rested with the welding company. SWEDAC, being the accreditation body, did not see it that way. A certification body is always responsible for its certifications, no matter what the timing. Our approach was to consider the six months signing as a surveillance of the welder. The certification body must control matters such as which person is qualified to sign the six months signature on a certificate, what will happen if the authorized person does not sign the certificate, what will happen if the welder leaves the welding company. The certification bodies had to develop procedures for handling these types of situations.

In the nondestructive testing area the *independence problem* was in focus. Unlike the welding area, the involved certification bodies shared our view: the independence of the examiner was a problem. Because the examiner in almost every certification case had some previous knowledge, the relation with the candidate for certification was definitively a threat to impartiality. The solution was to use both internal examiners and external examiners. When performing an examination, the internal examiner had to contact an external examiner, who selected a number of theoretical and practical tests. The internal examiner performed the written and the practical examination and scored them. The external examiner double-checked the written examinations as well as test pieces from the practical examination. Finally, the external examiner went through the internal examiner's scoring, and, if necessary, corrected it before approving the test result.

In the construction area, the problems were concentrated on the various normative documents. As we didn't know what the requirements were (how the candidates should be examined, with what type of examination, what should be the content of the examination or the extent of the examination), we were at a loss with respect to surveillance and recertification. This was our starting point in early 1995. The new legislation needed to be in force on July 1, 1995. The National Board of Housing Building and Planning called all interested parties to a meeting. Our message was that in order to accredit certification bodies, clearer requirements were needed both on the examination-certification procedure as well as factual requirements on the person to be certified. The training companies called for guidelines to be able to develop training courses that would meet the *requirements* in the regulations. The construction companies called for training courses for their personnel. The potential certification bodies looked confused, and the National Board of Housing Building and Planning said that they could do no more. As a result of this meeting, a working group was set up in order to solve the most urgent problems. After a couple of weeks, the task force came up with an interpretation document of the actual requirements on the certified person. The group had defined in detail the generic expressions of the regulation and the content of a written examination. The building authority accepted the guidance, and all different parties could continue with their part of the process.

The task force was unable to define the factual requirements for surveillance and recertification, but at least the accreditation process was initiated. We tried to stress the requirements for surveillance and recertification by requiring specific procedures from the certification bodies. During the first years of this scheme, the task force met with the National Board of Housing Building and Planning and SWEDAC several times. The objective was to reach a common understanding and implementation of the requirements. It was not a simple path; the discussions on the implementation of the requirements at times seemed endless.

The major lesson learned was that when the normative document was too *poor*, it created problems. During this period we also found weaknesses in our own regulations for certification bodies for persons. The weaknesses were most evident in the area of requirements on normative documents and examinations. By revising our own regulation, we were given a better tool to deal with this kind of situation.

Examples of formulations regarding normative documents and examinations in SWEDAC's previous regulations (The Swedish Board for Accreditation and Conformity Assessment's Regulations for Accredited Bodies that Certify Personnel, STAFS 1997:2) are as follows:

> If the normative document does not regulate the examination required, the accredited certification body shall regulate this in an interpretation document. These interpretations shall be accepted by all parties affected by the certification and be available to the public. If the normative document does not regulate the surveillance of the certified person during the certificate's validity, the validity of the certificate shall not be more than one year.
>
> Preferably the normative document shall be a national or international standard. If any other normative document is used this document should have been drafted together with the trade organization or equivalent. Such normative document should be replaced by a national or international standard as soon as such a document is available.
>
> A normative document for certification for personnel shall be sufficiently detailed and clearly framed so that two certification bodies, independently from each other, come to the same result. If the normative document used for certification is not clearly worded or is incomplete the certification body shall inform the person/body responsible for the normative document.
>
> Parameters that shall be regulated in a normative document for certification of personnel are:

- Basic education
- Practical experience
- Further training
- Examination (theoretical and/or practical)
- Other requirements (e.g., physical, authority requirements)
- Limited period of validity (1 to 5 years)
- Requirements during the period of validity of the certificate
- Surveillance of the certified person during the period of validity of the certificate
- Requirements for extension of a certificate's validity
- Reasons for withdrawal of certificates

By integrating these *rules* into our regulation, we gave ourselves much better communication tools in dealing with scheme owners and certification bodies.

One practical example is the applications from the technicians for ventilation who were approved in the *old school*. The authorities had not produced any transition rules, so the only route for the technicians was to contact an accredited certification body and apply for certification. Many argued that they should simply be given the new certificate, because they were previously approved. But neither the accreditation body SWEDAC nor the accredited certification body nor the national authority for building accepted this approval as a justification for competence under accreditation. The whole group of technicians had to undergo a new examination to justify their competence to achieve a certificate under accreditation.

In 1996 our discussions and the *new rules* produced a positive result. The certification criteria for persons responsible for quality and inspectors of function of ventilation systems were revised to better fit with accredited certification of persons. The new requirements encompassed requirements for both surveillance and recertification.

Certification Schemes and Normative Documents

During the second half of the 1990s SWEDAC received many inquiries regarding new areas for certification of persons, for example, cleaners, project managers, environmental auditors, and bracers. This meant a great deal of contacts with people and organizations who knew very little about certification and how a normative document for certification of persons should look. We almost spent more time discussing normative documents and the foundations of certification of persons than actually assessing certification bodies. The return in actual accreditations following all discussions and meetings was about 10%. Some of the areas were not suitable for certification. In some cases the interested party did not want to open up its scheme to other interested parties. And in other cases the population of potential certified people for a specific scheme was too small to interest any certification body to apply for accreditation.

It is typical for Sweden that a certification scheme for persons is developed by other organizations rather than by accredited certification bodies. This means that when a certification body applies for accreditation from SWEDAC for a specific scheme, we normally have no direct contact with the scheme owner. According to our experience, there is always some lack of compliance with the scheme and our requirements on the normative document. We have at times encountered communication problems with the scheme owner, maybe because the communication quite often is channeled through the applicant certification body. Because we have no day-to-day relations with the scheme owner, it has proved difficult to require changes in their normative document. The process to have an acceptable normative document for a scheme for certification could be as long as the actual accreditation process. Luckily, we have had accreditation processes for new certification schemes that we could process quite smoothly thanks to a well-informed scheme owner. This has been the case when the scheme owner has had an early contact with SWEDAC in order to understand accreditation and certification in general and especially the accreditation and certification process.

One of the major points in our information to scheme owners is that they should involve accredited certification bodies in the process of developing the scheme and the normative documents. The certification bodies are not there to give advice on the actual requirements on the certified person, but rather to contribute to the wording of the requirements and the process of verifying the requirements.

"The Most Perfect Scheme"

An interesting example of potential new schemes originated from a public debate about the costs for obtaining a driving license for cars. Certification of persons and driving licenses for cars, what a combination! This must be the ultimate certification scheme. SWEDAC was contacted by the national authority, asking for ideas on how accredited certification could be used in the area of driving schools and the issuance of driving licenses. We suggested involving accredited certification bodies for persons, the national authority and local authority for driving licenses and the police. Our main idea was that accredited certification bodies should examine and certify candidates for *competence in driving*. With the certificate in hand, the candidate could turn to the national authority and have the driving license issued.

The accredited certification body's responsibility includes the assessment of competence to obtain a driver's license; otherwise, the distribution of responsibility is unchanged. Figure 6.1 shows the operators and the primary contact/information routes described in SWEDAC's proposal for a national certification of driver's licenses. The driver's license of today would be substituted by a certificate issued by an accredited certification body.

Even though our suggestion was never implemented, the illustration mentioned previously shows a system enabling the use of accredited certification as the base for a decision in another forum.

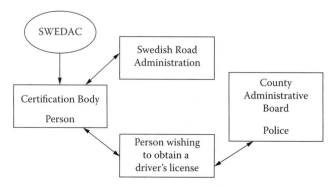

Figure 6.1 Certification process for driver's licenses.

The Value of a Scheme

A common question is the possible value of a foreign certification. Europe has employed a multilateral arrangement (MLA) for certification of persons for a number of years. The question, therefore, is not about the accreditation process or the certification process. What really counts, when it comes to certification of persons and acceptance of a specific certificate, is the certification scheme and the normative document. The buyer decides the value. My answer, as an accreditation body representative, is that the certificate is issued by a certification body accredited by a member of the European Co-operation for Accreditation MLA for certification of persons. This implies that the certification body and its procedures for certification are evaluated according to international standards by a competent organization and that you can rely on the certificate as if it had been issued by an accredited certification body in your country. It is, however, the questioner's job to find out how relevant the certification scheme is in his/her own country.

ISO/IEC 17024

Finally, I discuss the development of the ISO/IEC 17024. When ISO/CASCO suggested this project, we said "at last!" We had struggled for so long with the insufficient EN 45013 and truly needed an adequate tool in order to assess our certification bodies for persons in line with other types of certification bodies. It was an easy decision to join the ISO/IEC working group 17 to develop the new standard. Without going into detail on the actual development work, we are quite happy with the result.

The part dealing with development of certification schemes and normative documents, however, differs from the normal application in Sweden. As mentioned earlier, the 17024 implies that certification schemes and normative documents are developed within a certification body. On the contrary, the schemes are developed outside accredited certification bodies, and normally several certification bodies apply for accreditation for the same scheme. This is, however, not a problem. We are pleased with the section describing the development of certification schemes and normative documents. Finally, we have something to put in the hands of the interested parties and future scheme owners, and they have to comply with this part.

Another section I would like to emphasize in 17024 is the way of stating requirements on the quality management system of the certification body and the new function *scheme committee*. I am very much in favor of having ISO 9001:2000 as the criteria for the quality management system of the certification body. The operational requirements are set in 17024. All you have to do is apply the same requirements into your ISO 9001:2000 system. Why should conformity assessment bodies be the only organizations where ISO 9001:2000 is not relevant? This part has not been subject to any discussion or questioning in Sweden. The *old* certification bodies already had a quality management system that was functional with minor changes. And as a new certification body, it is quite evident that the best platform for the quality management system is ISO 9001:2000.

Speaking of the scheme committee, its structure and function are partly new to the certification bodies. They have all had representatives in their certification committee from interested parties. The new part is to keep the impartiality issues in the certification committee structure and leave the more technical issues to the scheme committee. The discussion we have had in Sweden is related to the many schemes with more than one accredited certification body. How should one prevent various scheme committees from developing different types of practices for the same scheme? The solution came from the IAF guidance on the implementation of ISO/IEC 17024, common scheme committees in which, for a specific scheme, all certification bodies, and interested parties are represented. This solution has been chosen by the certification bodies for almost all schemes where we have more than one accredited body. We have had very positive reactions—from the interested parties to the common scheme committees—and especially from the authorities concerned. This way, their communication with the accredited certification bodies and the remaining interested parties is clear and straight. It is much easier and quicker to affect the application of the certification scheme through the common scheme committee than going the formal way through regulations. Furthermore, a single focal point is less time consuming than dealing with each certification body separately. It is also a matter of justice. If an authority is involved as, for example, a scheme owner in accredited certification, it cannot be involved with one single certification body. Here, it is necessary to increase its involvement with all accredited certification bodies.

A general remark on the reaction of the new ISO/IEC 17024 is that it is not a big change for SWEDAC. We have had accredited certification bodies for persons since 1991. We have tried to update our regulations and our own procedures for accreditation in line with gained experience and development in other certification areas. We have been deeply involved with the development of both the ISO/IEC 17024 and the IAF guidance on the application of ISO/IEC 17024. Thanks to our insight during the development of the current standard and guidance, we were able to keep our accredited certification bodies and scheme owners informed about the future content. In that way our accredited bodies where quite fast in adopting the new requirements when they came into force.

The Near Future

Let me offer one final real-life example that, in a way, shows how the authorities and certification bodies' understanding for accreditation and certification have increased over the years, as compared to my previous example from the construction area. In connection with a workshop with accredited certification bodies for persons in May 2004, we were approached by a representative from one body. His concern was that the Swedish national authority for railways was purchasing a service for qualifying personnel in the railway sector without any connections with accreditations. His company and a few other accredited certification bodies had been invited to submit an offer on how to evaluate different kinds of occupations in the

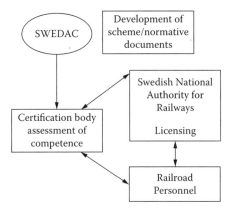

Figure 6.2 Swedish National Railroad certification process.

railway sector. They could also submit an offer for a system for registration of evaluated persons. He thought that it was a bit strange that accreditation was not mentioned in the invitation and wanted our comments on that. The very next day, I contacted the responsible person at the railway authority. We decided to have a meeting, and he supplied me with the relevant documentation. At the meeting, SWEDAC presented how a system with accredited certification bodies could fit into their system of approval of persons working in the railway sector. The solution we suggested left the final decision of approval of a specific person to the railway authority. Figure 6.2 shows the operators and the primary contact/information routes in SWEDAC's proposed system for certification and licensing of railroad personnel. The accredited certificate constitutes a necessary basis for the railroad authority when licensing approved personnel to work in the railroad sector. The normative documents can be developed by different *interested parties* depending on the scope of the scheme.

The benefits we pointed out were:

1. The Swedish National Rail Administration owns and controls the certification system. It can influence through specification of requirements, the joint technical committee, and as a technical assessor employed by SWEDAC.
2. The Swedish National Rail Administration issues qualifications.
3. Purchasing/agreements are not needed.
4. Costs are paid by respective interested party.

5. Certification bodies are responsible for their certifications.
6. Independent supervision of the system.
7. European harmonization.
8. Capacity and continuity.

The result of the meeting was that the railroad authority canceled the purchasing. During the autumn of 2004, they informed the interested parties of our proposal, and a decision was made to take the accreditation route. During 2005 a working group consisting of interested parties, the railway authority, and SWEDAC was set up to start the process. Certification bodies were involved in 2006 and the first accredited certifications are expected in 2007. The potential of the Swedish railway sector is enormous, partly because of the potential numbers of certified persons, partly due to the potential influence to other areas under public control.

Conclusions and Lessons Learned

Finally some advice based on more then 10 years of dealing with accreditation of certification bodies for persons:

1. The most important individual aspect for a certification scheme is the normative document. It must contain unambiguous requirements on the persons to be certified as well as the procedures for examination, surveillance, and recertification.
2. In order to achieve a normative document suitable for use under accreditation, the interested parties must be identified and given the opportunity to take part in the development of the certification scheme.
3. There must be a demand for a certified person. It does not matter if the potential certification market for a certain kind of certified person is enormous. If no one asks for or requires certified persons, people will not apply for certification.
4. If an *old system* of approval is replaced by an accredited certification, a transition period or an alternative way to a first-time certification must be considered.
5. Accreditation assures that the certification bodies involved are competent and impartial and that the scheme is accessible to all interested parties (certificants and certification bodies). Accreditation and certification involve a long-term process, in which the certification body and the certified company each are evaluated on

a continuous basis. Certificates under accreditation are accepted internationally. (Europe has a MLA for certification of persons.)

Notes

1. Scheme: ISO/IEC 17024, 3.4 certification scheme specific certification requirements related to specified categories of persons to which the same particular standards and rules, and the same procedures apply.
2. Sweden has been a member of the European Union since January 1, 1995.
3. Examiner: ISO/IEC 17024, 3.10 examiner person with relevant technical and personal qualifications, competent to conduct and/or score an examination

References

CEN/CENELEC, The Joint European Standards Institution. (1989). European Standard EN 45013 General criteria for certification bodies operating certification of personnel. Central Secretariat: rue Bréderode 2, B-1000 Brussels. Withdrawn October 2003.

International Organization for Standardization. (2003). *International standard ISO/IEC 17024 conformity assessment—general requirements for bodies operating certification of persons.* Geneva, Switzerland: Author.

The Swedish Board for Accreditation and Conformity Assessment. (1997). The Swedish Board for Accreditation and Conformity Assessment's regulations for accredited bodies that certify personnel. *STAFS* (2).

7

Accreditation in the Certification and Licensing Sector of the Testing Industry
Current Status and Future Possibilities

Sharon M. Goldsmith
Goldsmith International SP
Plexus Consulting Group LLC

Michael Rosenfeld
Rosenfeld & Associates

Introduction

Purpose

There is a broad area of testing guidance that addresses testing in general. There is also a rapidly growing area of guidance for testing that is used to make certification and licensing decisions (National Organization for Competency Assurance, 2003; International Organization for Standardization [ISO], 2003; American National Standards Institute, 2004). Accreditation is becoming an increasingly popular tool to promote quality across diverse industries (Joint Commission, n.d.; Council for Higher Education Accreditation, 2006). This chapter explores the potential utility of accreditation as a quality driver and a quality indicator for the testing industry. The focus of this chapter is on that segment of the testing industry that supports personnel certification and licensure.

Uses of Accreditation: Principles and Definitions

Accreditation has increasingly become a way to identify individuals, agencies, or products that can demonstrate that they conform to an accepted set of quality standards. However, there is no universally accepted definition of accreditation or the processes that constitute accreditation. According to the Canadian Information Centre for International Credentials (2003, p. 2) accreditation is "A process of quality assurance through which accredited status is granted ... by responsible authorities. It means that standards ... established by professional authorities have been met ... The process usually includes self-assessment by the program under review ... and on-site visits by qualified, external reviewers from government and/or nongovernmental agencies. A program's accreditation status is normally subject to periodic review and may be withdrawn by relevant professional authorities." An accrediting body is further defined by the Canadian Centre for International Credentials (2003, p. 2) as "The authority that is acknowledged as having the responsibility of granting accreditation ... Accrediting bodies can be (but are not necessarily) mandated by legislation or by regulatory bodies and can consist of government representatives, stakeholder representatives, external academic experts, and professional regulatory bodies."

Accreditation generally consists of four elements:

- **Standards** that are reflective of appropriate practices in an industry and indicate that if met an organization has the ability to operate in an ethical, legal, and technically sound manner that protects the public interest. As defined in the ISO/IEC Guide 2: 1996 Standardization and related activities—General vocabulary, "a 'standard' is a document, established by consensus and approved by a recognized body, that provides, for common and repeated use, rules, guidelines or characteristics for activities or their results, aimed at the achievement of the optimum degree of order in a given context" (ISO/IEC, 1996, p. 1).
- **Assessment and verification systems** that assess and make judgments regarding whether or not any applicant for accreditation complies with the accreditation standards and demonstrates that it has the capacity to continue to comply with the standards.
- **Independent decision making authority:** Assessment and verification systems must be operated by an accreditation agency that is independent of the organization being assessed and must be recognized by government, industry, or the public as having the authority to make accreditation decisions. This authority is generally representative of, or seeks input from, individuals and organizations that have an interest in the activities of the agency being accredited.

- **Enforcement:** Accreditation is generally granted for a limited period of time. Entities wishing to continue to be accredited must formally demonstrate their continued compliance with accreditation standards. Additionally, accreditation authorities generally have mechanisms to withdraw accreditation or impose other sanctions on agencies that violate accreditation requirements.

Accreditation offers many benefits to organizations seeking accreditation and to the general public. According to the International Accreditation Forum Inc. (IAF), an organization comprised of internationally recognized accreditation agencies in 44 countries, "accreditation reduces risk for business and its customers by assuring them that accredited bodies are competent to carry out the work they undertake within the scope of their accreditation" (International Accreditation Forum, 2004, p. 2).

Certification is similar to accreditation in that certification refers to processes that also publicly recognize entities that have demonstrated compliance to a defined set of standards. According to the International Organization for Standardization (ISO) "in some countries, bodies which verify conformity of products, processes, services or systems to specified standards are called 'certification bodies', in other countries 'registration bodies' and in still others 'assessment bodies'" (ISO/IEC Guide 61:1996). In the United States the term certification is generally applied to bodies that assess the quality of personnel or products, whereas accreditation is the term generally applied to assessment of systems, facilities, or processes.

Why Consider Accreditation as a Quality Improvement Tool for the Testing Industry?

As is evident in the following sections of this chapter, accreditation as a means to foster quality improvement is rapidly growing in popularity across diverse industries. Additionally, accreditation is being increasingly relied upon by the public, the government, and consumers to make judgments about the quality of specific organizations. It is reasonable therefore to consider whether the concept of accreditation could be useful in addressing issues related to quality in that portion of the testing industry involved in the areas of licensing and certification.

In summary, it is the growing popularity of accreditation from both industry and consumer perspectives that frames the environmental context within which this chapter was conceived. One question that needs to be addressed is: Why accreditation? Furthermore, given the success of accreditation in other venues, the question also needs to be: Why not accreditation?

Accreditation as a Quality Driver and Indicator of Quality: Examples from Other Industries

Several industries have incorporated the use of accreditation programs. For example, higher education programs are routinely accredited by one of over 62 nationally recognized specialty accrediting agencies in such fields as law, medicine, engineering, nursing, architecture, and accounting. Additionally several smaller accrediting organizations exist in such fields as audiology, clinical lab science, and interior design. Such accreditation is highly sought after by the institutions of higher education and is relied on by students and their families when making decisions about what educational programs to consider. In fact, accreditation in higher education is considered so important that the U.S. Department of Education (USDE) requires that higher education programs be accredited by a USDE approved accreditation agency in order to participate in federally subsidized student loan programs, receive federal grants, or participate in certain research projects. The USDE Office of Post Secondary Education has established a comprehensive set of standards for the design and operation of accrediting agencies. In this particular instance there is a formal system to *accredit the accreditors* (U.S. Department of Education, n.d., retrieved January 10, 2005).

A parallel higher education accreditation agency approval system exists in the private sector. A private agency, the Council on Higher Education Accreditation (CHEA), develops and implements standards relating to good practice in accreditation and recognizes those accrediting agencies that demonstrate compliance with these standards (Council for Higher Education Accreditation, 2006).

Recognition as an accrediting agency through USDE and CHEA is considered valuable in an environment where the number of accrediting agencies is proliferating and multiple agencies compete to accredit academic programs in the same discipline. For example, entry level nursing programs can now chose between two accreditors. Some disciplines, including law or business, have multiple accreditation agencies that vary in quality and reputation. State licensure boards in certain disciplines require that individuals be graduates of accredited programs as a condition of eligibility to register for state licensing examinations.

Medicine has long relied on accreditation as a public demonstration of quality. Accreditation of medical schools began in the United States in 1942 (National Health Policy Research Institute, 2003). Private accreditation agencies such as the Joint Commission (previously the Joint Commission on Accreditation of Healthcare Organizations [JCAHO]) accredit hospitals and related health care facilities that meet rigorous and comprehensive standards of quality for all aspects of their operation including facility cleanliness, calibration of medical laboratory instruments, quality of personnel, error reporting and maintaining programs of self evaluation, and continuing quality improvement (Joint Commission, n.d.). As in the academic environment, the public has come to rely on accreditation to make judgments about a health care facility. The use of outside agencies on accreditation has major impact for the health care community. Several government agencies such as the Center for Medicare and Medicaid Services (CMS) rely on accreditation as an indicator of quality and require Joint Commission accreditation or state approval for government funding (Sprague, 2005). Only accredited facilities are eligible to be reimbursed by certain private insurers or to participate in government payment programs such as Medicare and Medicaid. These factors have contributed to the fact that a 2003 American Hospital Association study found that 80% of U.S. health care facilities are Joint Commission accredited (Sprague, 2005).

The strong reliance on accreditation of health care facilities has encouraged many of the organizations that supply goods and services to health care facilities, like medical laboratories, to seek accreditation. Some of these accreditation systems have been developed and are managed within specific industries. For example, the College of Clinical Pathologists (CAP), a private organization whose members are physicians, accredits laboratories. According to its Web site:

The CAP Laboratory Accreditation Program meets the needs of a variety of laboratory settings from complex university medical centers to physician office laboratories. The program also covers a complete array of disciplines and testing procedures. The Centers for Medicare & Medicaid

Services (CMS) has granted the CAP Laboratory Accreditation Program deeming authority. [Deeming authority means that a governmental entity officially recognizes an accreditation agency's authority to make accreditation decisions that will be used for governmental purposes.] It is also recognized by the Joint Commission on Accreditation of Healthcare Organizations (JCAHO), and can be used to meet state requirements. (College of American Pathologists, 2000)

Other laboratory accreditation programs are based on standards created by independent third parties. For example, many medical laboratories have chosen to seek accreditation based on an international standard for laboratories (ISO/IEC 17025) that have been developed by the International Organization for Standardization (ISO) (American Association for Laboratory Accreditation, 2005).

In addition to industry-specific accreditation standards, there are generic, *industry-neutral* standards for maintaining quality systems that cross industries. One of the best known examples is ISO 9000 and ISO 9001:2000. These standards do not address specific industries, products, or services but rather they look at the processes used within an industry and a specific organization to assure quality. The requirements also address the manner in which these quality management systems are documented, implemented, evaluated, and modified (Perry Johnson Inc., 2006).

Joint Commission International (JCI), a division of the Joint Commission, grants accreditation to a rapidly growing number of eligible healthcare institutions operating outside the United States. The strong interest of the international community to seek accreditation suggests the growing universal appeal of accreditation.

Although medicine has relied on accreditation for a long time, several industries are newly developing accreditation programs. These programs are being developed in response to the need for greater accountability and the need for increased consumer confidence in an industry. One example is the new accreditation program that will be implemented by the Land Trust Alliance (LTA) in 2008. The LTA is a national organization that represents the interests of 1,500 land trusts—local, state, regional, and national groups that promote environmental conservation through land conservation. The rationale for developing the accreditation program and its goals are summarized in a press release issued by the Land Trust Alliance on September 29, 2005 (Land Trust Alliance, 2005). The press release reads in part:

In a year marked by Congressional scrutiny of charities and conservation land transactions ... LTA is leading the nonprofit community by
establishing a private-sector accreditation program that will recognize
the high standards employed by land trusts across the nation, will maintain the public trust in voluntary land conservation ... Accreditation
will require land trusts to adopt Land Trust Standards and Practices,
the ethical and technical guidelines for the responsible operation of land
trusts ... Charitable donors, government officials and the public will look
to accreditation as a seal of approval in land conservation. Eighty percent
of the nation's land trusts plan to participate in the program and seek
accreditation.

In summary, accreditation as generally defined and used in the
United States as a voluntary system. In some industries, however,
such as health care and higher education, accreditation is in practice a nonvoluntary system in that accreditation is required by most
third-party payers and government agencies and is generally recognized and even demanded by the public as a minimum benchmark of
acceptable quality.

Accreditation in the Testing Industry:
Existing Standards of Good Practice

Although there are many standards for testing (Code of Fair Test Practices [Joint Committee on Testing Practices, 2002]; Standards for Educational and Psychological Testing [American Educational Research
Association, American Psychological Association, & National Council
Measurement in Education, 1999]) there are only a few accreditation
programs that apply to the testing industry. These accreditation programs focus on certification bodies. Certification bodies are increasingly relying on test contractors to perform many of the functions of
a certification program. The same standards apply whether the certification body performs these functions using internal staff or sub-
contracts these functions. The certification body must provide direct
oversight of all activities that any contractor performs and must document that relevant accreditation requirements are being met.

The increased interest in certification bodies to seek accreditation
has fueled the interest of the testing community in understanding
those accreditation requirements that are applicable to the activities
they perform on behalf of their clients. The accreditation review process requires accreditation assessors to review test contractor procedures and, if necessary, interview test contractors and visit their

facilities and test sites. The strong involvement of the test contractor in the accreditation process has raised the question of whether there should be an accreditation process that in addition to accrediting certification bodies also accredits test contractors. The question of whether this can be the same accreditation process, whether this should be a parallel process or, more critically, whether this concept is even useful or feasible is discussed later in the chapter.

There are two generic, that is, industry-neutral accreditation programs available to personnel certification agencies. One of these programs, operated by the National Commission for Certification Accreditation (NCCA), an independent arm of the National Organization for Competency Assurance (NOCA), offers an accreditation program for personnel certification agencies. NOCA is a voluntary membership organization comprised of personnel certification agencies and other agencies, customer groups, governmental agencies, testing agencies, and other individuals who are interested in personnel certification.

The NCCA Standards for the Accreditation of Certification Programs (NOCA, 2003) is composed of 21 standards that are grouped into five sections: (1) Purpose, governance, and Resources, (2) Responsibilities to Stakeholders, (3) Assessment Instruments, (4) Recertification, and (5) Maintaining Accreditation. The section on assessment instruments contains nine standards that address issues related to test construction, use and interpretation of assessment test scores, job analysis, setting passing scores, reporting of scores, security, and record keeping.

Another program is offered through the American National Standards Institute (ANSI), which is a private nonprofit organization whose members are companies that have an interest in standards development and accreditation. ANSI "... administers and coordinates U.S. voluntary standardization and conformity assessment activities" (ANSI, 2002, p. 2). ANSI accredits personnel certification bodies using international standard 17024 developed by the International Organization for Standardization (ISO) and the International Electrotechnical Commission (IEC). Like NCCA, ISO/IEC 17024 specifies a broad set of requirements that certification bodies must meet in order to qualify for accreditation. ISO/IEC 17024 contains 47 standards or substandards that detail requirements for test administration, scoring, analysis, test construction, records maintenance, appeals of test results, test validity and reliability, test security, confidentiality, qualifications for staff, subcontracting, and management systems. These standards are grouped under three major headings: Requirements for certification bodies, Requirements for persons employed or contracted by a certification body, and the Certification Process. *Guidelines on Psychometric Requirements for ANSI Accreditation, ANSI-PCAC-GI-502* interpret accreditation requirements contained in Section 4.3.6 of ISO/IEC 17024 as they apply to testing "methodologies, procedures, types of analyses and how they are applied" (ANSI, 2004). As explained in an article written by the ANSI Personnel Certification Program Director, "While the criteria under which a personnel certification program is evaluated are set forth in ISO/IEC 17024, the actual accreditation process is based on procedures contained within the international standard for *General requirements for accreditation bodies accrediting conformity assessment bodies—ISO/IEC 17011*" (Swift, 2005a, p. 2).

ISO/IEC 17011 outlines an accreditation system that includes the process for evaluating documentation provided to determine if the certification body meets each of the requirements of the specific standard, and validating information presented during an on-site audit conducted by trained assessors. "The findings of the assessors are presented to an oversight body comprised of nationally recognized experts in certification and test development and a broad representation of industries and the public. ... Upon approval, accredited bodies must also engage in an annual reporting process that details any changes to their program; in some cases an additional on-site visit is required" (Swift, 2005a p. 2).

In addition to industry-neutral accreditation standards, there are standards that have been developed for accrediting or otherwise approving certification programs in specific industries, for example, specialty nursing and specialty psychology occupations. Perhaps the most comprehensive standards that certification programs must adhere to in regard to testing are the standards for assessment contained in the Conference for Food Protection: Standards for Accreditation of Food Protection Manager Certification Programs (2004). The Conference for Food Protection standards provide explicit accreditation requirements relating to assessment and program operations.

There are also approval mechanisms specifically directed toward testing companies. These approval mechanisms closely approximate accreditation programs. One of these that is industry neutral is developed by Buros Institute for Assessment Consultation and Outreach (BIACO). These standards focus on the psychometric principles involved in good test construction in order to assure that tests are fair, reliable, and valid. The Buros Institute has developed standards for proprietory testing programs. "BIACO uses these standards to review the quality of proprietory testing programs" and will accredit tests that meet these standards (Buros, 2006, p. ii). The standards address the structure and resources of the testing program, examination development, examination administration, scoring and score interpretation, exam security, and responsibilities to examinees and the public (Buros, 2006, p. iv).

The Buros standards focus on specific test forms, not the general operations of the testing company or certification agency and its multiple functions. As discussed later in this chapter, the development of test forms is only one part of what testing companies do.

Another example of a formal approval mechanism developed with specific industry needs in mind is managed by the Association of Real Estate License Law Officials (ARELLO). ARELLO, through its Examination Certification Council, approves real estate broker and salesperson testing programs that can demonstrate compliance with a published set of standards. Approval is good for five years (ARELLO, 2002). The test contractor is not approved in its entirety; what is being approved is its real estate testing program. The processes and procedures are approved only as they address the real estate testing program. The activities performed for other clients are not considered or evaluated.

Many governmental agencies are beginning to consider relying on accreditation as an indicator of quality in certification programs and to help both government agencies and consumers make informed decisions (Goldsmith, 2006). A recent example is the 2006 U.S. Department of Defense Directive 8570.1 that requires "all individuals who perform certain Information Assurance functions to be certified by certification bodies that have been independently accredited by a third party" (ANSI, 2006, p. 1).

The International Organization on Standardization (ISO) over the objections of the U.S. testing community and of ANSI, the U.S. representative to ISO, has recently convened an international working group to create an international standard for Psychological Assessment Services. This international work group will "standardize requirements for psychological assessment services with reference to assessment planning, selection, integration, implementation, evaluation, and interpretation, and the qualifications of individuals participating in the assessment process" (ISO, 2006, p. 1). If approved, this standard will become the basis for a voluntary, international system to accredit test vendors directly. However, negotiations began in March 2007 and issues such as the scope of the standard are still being negotiated, The ANSI authorized U.S. Technical Advisory Group is being coordinated by the Association of Test Publishers.

Approach to Exploring the Usefulness of Accreditation as a Way of Promoting Quality in the Certification and Licensing Sector of the Testing Industry

Because this is a relatively new area, our approach was to interview experts in the field of testing, certification, and accreditation to seek their opinions regarding the value of accrediting testing contractors. Individuals interviewed included test contractors, educators responsible for large-scale testing programs, state licensing authorities, accrediting agencies, and personnel certification agencies. All the interviews were conducted in person or over the telephone. Twenty individuals were approached to be interviewed; fifteen actually participated in the interview process. Of the 15 individuals interviewed, 10 represented the testing community, 6 represented the certification and licensing community, and 3 represented the accreditation community. These numbers add up to more than 15 because several individuals had experience in and represented multiple perspectives. Some individuals reported that they were voicing their own beliefs; others indicated that they had spoken to colleagues to assess their opinions as well. Individuals were sent a written protocol outlining the purpose and content of the interview.

The intent of the interviews was not to conduct a scientific survey of people's views but rather to seek an informal sense of the thinking of a few individuals familiar with issues regarding accreditation and/or testing that might supplement and help organize the authors' thinking regarding some of these issues. Individuals were told that their opinions would not be individually identified and only the broad industry groupings they represent would be provided in order to maintain their anonymity. The authors sincerely thank everyone who helped us think through the issues by sharing their advice, expertise, and time. Although we have attempted to represent some of the opinions we heard, the views expressed in this chapter are those of the authors.

TABLE 7.1   Value and Uses of Test Industry Accreditation—Interview Questions

1. Do you think that an accreditation program to accredit testing contractors (contractors) is a good idea?

    From the perspective of a certification agency?
    From the perspective of a testing contractor?
    What are some of the reasons to support this idea?
    What are some areas of concern or reasons for not supporting this idea?

2. Are there any trends in the testing industry or in related industries that you think impact on the issue of potentially developing an accreditation system? If so, what are they?

3. If you think that the concept of an accreditation program is a good idea, what should the goals be for this type of accreditation program? What should accreditation encompass?

4. Do any relevant standards already exist that can be utilized in an accreditation program? If so, what are they? Do new ones need to be developed?

5. What entities are best suited to develop or assemble standards, administer the accreditation process, and grant accreditation?

6. What do you see as the practical constraints or limitations if this idea were to be pursued and implemented?

Interviews generally lasted about one half hour and covered the questions outlined in Table 7.1.

Advantages of an Accreditation System for the Testing Industry

Accreditation could benefit the testing industry in several ways. From the perspective of the consumer it might give some comfort to a client that they were hiring a competent contractor by presumably providing assurance that the accredited contractor would always use qualified staff, have sufficient resources, and follow professional testing standards for developing and administering fair, reliable, and valid exams as well as appropriate collateral materials. It would give individuals who are responsible for hiring test contractors some indication of the competence of organizations—particularly those individuals who were not very knowledgeable about the testing. It would allow the consumers of testing services to make more informed choices when hiring test contractors.

From the perspective of the industry, accreditation standards might provide additional guidance to the testing industry—beyond existing standards of what constitutes good practice. For example,

ISO/IEC 17024 includes requirements for management systems as well as requirements that apply to test construction and use. Accreditation standards could be used to develop training or instructional materials used in testing companies or in university-based programs in tests and measurements.

As evidenced in other industries, accreditation might provide a competitive business advantage to accredited test contractors. It would demonstrate to regulatory authorities the industry's interest in policing itself and might forestall any governmental or third-party regulation being imposed on the industry. Accreditation against international standards, for example standards that might be parallel to ISO/IEC 17024, could assist certification agencies and contractors who wish to globalize their operations. Many of the large test contractors are already international in scope and might benefit from being able to demonstrate that their procedures comply with international standards and that quality practices are consistently applied regardless of what country tests are being offered in.

Challenges in Creating an Accreditation System for the Certification and Licensing Sector of the Testing Industry

Individuals interviewed expressed strong concerns regarding the necessity and feasibility of an accreditation program for the certification and licensing sector of the testing industry. No one that participated in the interviews regardless of whether they represented the testing industry, the certification industry, or the state regulatory perspective, was strongly in favor of accrediting test contractors. Some interviewees representing the test industry perspective felt it was a little awkward that the industry did not have an accreditation program when related industries, such as personnel certification, did. Interviewees across all industries, however, commented on the fact that there were no blatant problems in the testing community. Neither private certification programs nor state agencies reported any difficulties finding competent contractors. The industry has worked with the sets of guidelines described previously for many years. They appear to be working well and although there have been isolated instances of errors in scoring and breaches in security there have not been any reported wide-scale abuses within the industry.

The testing industry in the United States, for good or for bad, is dominated by a few large-scale test contractors. These large-scale contractors have the technical abilities, the resources, and the interest to implement best practices in testing and the related functions they perform.

Even in the absence of formal accreditation systems there are several major factors that motivate test contractors to adopt and follow best practices in testing. These include:

1. Severe competition among contractors for clients resulting in strong interest in providing quality services.
2. Informal peer oversight as a result of the relatively small and highly interpersonal nature of the testing community where successes and problems quickly become very public.
3. High legal exposure inherent in testing from court cases regarding perceived bias and discrimination as well as errors in scoring and equating.
4. Strong fiduciary interest in protecting the security of tests and public oversight of testing issues through the media and consumer advocacy groups.

Although contractors make errors from time to time, neither voluntary guidelines of good practice or accreditation will eliminate the occasional error, just as accreditation of medical teaching facilities and heath care delivery facilities has failed to eradicate all instances of malpractice or medical error.

Interviewees commented on the difficulty, cost, and time necessary to set up a mechanism for managing and overseeing accreditation particularly without strong backing from industry or the public to do so. Some representatives of the testing industry were concerned about the additional costs inherent in participating in an accreditation program through accreditation application, maintenance, and renewal fees. Test users expressed concern that additional costs would be passed along to them and/or test candidates.

A single rigid model of accreditation that required that test contractors be accredited in all the services that a test contractor might possibly provide may not be feasible due to the broad scope of services that test contractors provide and the diverse types of services they provide to different clients. Testing agencies perform multiple functions. The types of functions are dictated by the needs of the client and the scope of these functions is often dictated by what the client is willing or able to pay. The range of services test contractors may provide can include developing candidate guides, developing training and test preparation material, test registration, designing and implementing accommodations for individuals with disabilities, test development, test translation, security procedures, test administration, scoring and reporting test scores, psychometric analysis and monitoring, recertification, promotion and marketing, and association management. Not all contractors perform all of these functions.

Issues to Be Resolved in Order to Develop an Accreditation System for the Testing Industry

The Joint Committee on Testing Practices discussed that if accreditation becomes desirable or necessary new models of accreditation will need to be developed to meet the unique needs of the industry. Regardless of the impetuses for accreditation, the following questions need to be resolved before accreditation could be a feasible and useful tool to foster quality practices in the testing industry:

1. What should be the goals of an accreditation program? What specifically should it be designed to accomplish?
2. What kind of entity are we accrediting and what are we accrediting this entity to do? What would accreditation mean to the industry and to the public?
3. Are there some activities performed by testing contractors that may be more amenable to accreditation than others?
4. Would there need to be reaccreditation systems? If so, what is a useful length of accreditation? Would that time period be the same for contractors working in all industries?
5. Should accreditation be voluntary? What will the impact of accreditation be on test contractors that chose not to become accredited or for other reasons cannot successfully participate in the accreditation process?
6. Are there a sufficient number of testing companies to warrant an accreditation program? What level of participation among test contractors would be necessary to sustain an accreditation program?
7. Can/should one create a *one size fits all* accreditation that covers the entire range of functions that a test contractor might perform, or must there be a specialized (compartmentalized) accreditation system with specific standards for each function? In the latter case would testing contractors then potentially earn multiple accreditations?
8. How would a system of accreditation be flexible enough to handle expansion or reduction of scope of test contractor operations?
9. What are the implications of accreditation on a contractor's business operation? Does it mean that an accredited contractor will

chose not to do certain things or be prevented under the conditions of accreditation from doing certain things? Does it mean that the contractor will apply a certain level of quality to all operations that are accredited and will refuse jobs from clients that are unwilling or unable to pay for this level of service?

10. What if a certification program has testing functions divided among multiple contractors? In this circumstance it appears that a modular approach would seem to be the one that would be most workable and in fact would provide maximum flexibility to certification agencies that could then pick and chose among contractors who have particular areas of expertise.

11. If a broad accreditation is developed, how would that impact the small test contractor who runs a one person shop and doesn't provide the scope of possible services? It appears that a broad accreditation may penalize the small contractor. A modular approach would allow contractors to specialize in specific areas and seek accreditation in only those specific areas. Contractors would, however, need to be careful not to misrepresent the scope of their accreditation.

12. Given the volatility of the industry, what happens to accreditation when companies merge, or are acquired by another contractor or business?

13. How would accredited contractor operations be monitored? Many accreditation systems have a code of ethics that accredited agencies must abide by. Most accreditation systems include safeguards to protect the integrity of the accreditation process whereby accredited agencies that subsequently violate accreditation standards or ethical principles can be identified, investigated and, if appropriate, be subject to mechanisms that can publicly reprimand the agency and remove its accredited status. If a contractor violates an accreditation standard for one aspect of its business would all other aspects of its business also be subject to reprimand?

14. What entity(ies) is best suited to develop standards, administer the accreditation system, and award accreditation? Is it the same entity or should it be multiple entities. Are we looking at a national system or an international system? What are the pros and cons of each model?

Possible Alternatives to Accreditation

There may be other less expensive and more effective alternatives that can be implemented to continue to promote quality in the testing industry and to educate the public that participate in testing programs and agencies that hire testing contractors as to what to look for regarding a quality program.

The Joint Committee on Testing Practices discussed the need for additional public education in the area of testing (JCTP, 2005). Such education could include further guidance on the uses and abuses of testing, what to look for in selecting tests, and how to hire a testing contractor. Such guidance can assist the public in recognizing quality testing practices and test contractors.

Test users should be encouraged to informally share information with each other regarding positive as well as negative experiences with testing contractors. One positive step in this direction has been the increased participation of test users at meetings with developers of tests in industry sponsored conferences such as the annual Association of Test Publishers (ATP) meeting. Such venues, as well as those available through NOCA, ANSI, and CLEAR (Council on Licensure, Enforcement, and Regulation) can create structured opportunities for all parties to discuss concerns and solutions.

Testing contractors might also benefit from participating in workshops such as those offered through the ANSI to learn more about the international accreditation standards (ISO/IEC 17024) as they relate to quality standards for the development and use of tests.

National Federations of Licensing Boards can develop guidelines and training opportunities for their member licensing boards on selecting and working with test contractors. Guidelines could be developed and sent to licensing or certifying boards that provide information on how to audit their program.

Although it may not be viable to accredit a testing contractor due to the broad and diverse range of services they perform, several interviewees suggested that it might be useful to develop personnel certification requirements for key individuals that perform specific functions in testing companies such as pyschometricians.

Conclusions and Lessons Learned

Based on the interviews, there does not appear to be a strong demand to develop an accreditation program for the testing industry. The industry is voluntarily using several sets of testing standards or guidelines. Related accreditation systems in certification are encouraging the use of quality practices in the testing industry. In addition, wide-scale problems do not appear to exist and interviewees reported no difficulty in finding competent contractors.

Given the number and complexity of the questions outlined earlier, developing an accreditation system for the testing industry, even if it was clear that one was needed, would be a difficult, time-consuming, and expensive process. At this time, it appears that accreditation would not provide sufficient added value to warrant the resources necessary to create and implement such programs.

However, additional guidance materials describing how to contract with, monitor, and evaluate the services provided by test contractors would be valuable tools for newly developing licensing and certification agencies and likely improve the quality of their products and services.

Because of the experiences of several other industries, market forces and public or governmental demand may create a need for more formal and industry-targeted accreditation systems in the future. If this were to occur, the answers to the questions raised in this chapter will contribute to the formulation of viable accreditation models and programs that meet the needs of the testing industry and its consumers.

References

American Association for Laboratory Accreditation. (2005). *Understanding ISO/IEC 17025:2005, Specific applications of the standards*, http://www.a2la.org/faq./list17025faqs.cfm

American Educational Research Association, American Psychological Association, & National Council on Measurement in Education. (1999). *Standards for educational and psychological testing.* Washington, DC: American Educational Research Association.

American National Standards Institute. (February, 2006). Draft press release: U.S. Department of Defense to require compliance with international standard for contractors and DOD personnel. Washington, DC: American National Standards Institute.

American National Standards Institute. (n.d.). *Hallmarks of Distinction, ANSI Personnel Certification Accreditation Program*, Washington, DC: Author.

American National Standards Institute. (2004). *Guidelines on psychometric requirements for ANSI accreditation*, ANSI-PCAC-GI-502. Washington, DC: Author.

American National Standards Institute. (2002). *National conformity assessment principles for the United States*. Washington, DC: Author.

Association of Real Estate License Law Officials. (2002). ARELLO Examination Certification Council. *Guidelines for certification.* Toronto, Canada: Association of Real Estate License Law Officials.

Association of Specialized and Professional Accreditors. (2005). *ASPA: Association of Specialized and Professional Accreditors Membership Directory 2005-2006.* Chicago: Author.

Buros Institute for Assessment, Consultation and Outreach. (2006). *BIACO standards for proprietory testing,* http://www.unl.edu/buros/biaco/pdf/standards03.pdf, pp. ii–xxii.

Canadian Information Centre for International Credentials. (2003). *Guide to terminology usage in the field of credentials recognition and mobility in English Canada.* Canadian Information Centre for International Credentials. Retrieved May 1, 2005, from http://www.cicic.ca/pub/guide/Guide, pp. 1–16.

College of American Pathologists. (2000). *Standards for Laboratory Accreditation 2000 Edition.* Retrieved December 1, 2005, from http://www.cap.org/apps/docs/laboratory_accreditation/standards/standards.html

Conference for Food Protection. (2004). *Conference for Food Protection: Standards for accreditation of food protection manager certification programs.* Gilroy, CA: Author. Available online at http://www.foodprotect.org./pdf/standards.pdf

Council for Higher Education Accreditation. (2006). *Recognition of accrediting organizations policy and procedures.* Available at http://www.chea.org

Goldsmith, S. (2006, May). *Uses of accreditation.* Presentation to Land Trust Accreditation Commission, New Paltz, NY.

Goldsmith, S. (2005). *Defining commonly confused terms in regulating personnel.* Unpublished manuscript.

Goldsmith, S., Swift, R., & Briggman, C. (2003). *Accreditation of personnel certification agencies: The other component to international quality assurance.* Presentation to Eleventh Annual Forum on the Globalization of Higher Education, Washington, DC (proceedings in press).

International Accreditation Forum (2004, February). *IAF guidance on application of ISO/IEC 17024- 2003 IAF GD 24-2004,* Issue I. Available at http://www.iaf.nu

International Organization for Standardization. (1996). *ISO/IEC Guide 2: Standardization and related activities—general vocabulary.* Geneva, Switzerland: Author.

International Organization for Standardization. (1996). *ISO/IEC Guide 61:1996 General requirements for assessment and accreditation of certification/registration bodies.* Geneva, Switzerland: Author.

International Organization for Standardization. (2003). *International standard ISO/IEC 17024 conformity assessment—general requirements for bodies operating certification of persons.* Geneva, Switzerland: Author.

International Organization for Standardization. (2006). *New work item proposal: Psychological assessment services.* Geneva, Switzerland: Author.

Joint Commission. (n.d.). *Frequently asked questions about hospital accreditation.* Retrieved May 1, 2006, from http://www.jointcommission.org/AccreditationPrograms/Hospitals/faqs.htm

Joint Committee on Testing Practices. (2000). *Bylaws.* Retrieved December 31, 2005, from www.pa.org/science/jctpweb.html

Joint Committee on Testing Practices. (2002). *Code of fair testing practices in education.* Retrieved December 31, 2005, from www.apa.org/science/jctpweb.html

Land Trust Alliance. (2005). Press release, September 29, 2005. *Land Trust Alliance approves strong accreditation program.* Washington DC: Land Trust Alliance. Available at http://www.lta.org/newsroom/

National Health Policy Research Institutes. (2003). *Taiwan Medical Accreditation Council origin.* Retrieved November 19, 2003, from www.nhri.org.tw/nhri-org/mc/mainl.html

National Council of State Boards of Nursing. (1997). *National Council of State Boards of Nursing position paper on approval and accreditation: definition and usage.* Chicago: Author.

National Organization for Competency Assurance. (2003). *National Commission for Certifying Agencies standards for the accreditation of certification programs.* Washington, DC: Author.

Nelson, D. (2005, June). *Assessing fairness validity and reliability of oral, written and skill assessments.* Presentation to IAF TC Working Group on ISO Personnel Certification Accreditation ISO/IEC 17024, Washington, DC.

Perry Johnson Inc. (2006). *Description of ISO standards.* Retrieved May 1, 2006, from http: www.pji.com/iso.standards.htm.

Sprague, L. (2005, May). *Hospital oversight in medicare: Accreditation and deeming quality.* (Issue Brief No. 802, pp. 1–15). Health Policy Forum. Washington, DC: George Washington University.

Swift, R. (2005a, October). Global accreditation: Building worldwide confidence in personnel certification programs. *ANSI Reporter* (pp. 5–6). Washington, DC: American National Standards Institute.

Swift, R. (2005b). Accreditation of personnel certification agencies: A component to international quality assurance. *ASAE Global Link* (pp. 1–3). Washington, DC: American Society of Association Executives.

U.S. Department of Education. (n.d.). *Accreditation in the United States.* Retrieved January 10, 2005, from http://www.ed.gov/adminis./accred/accreditation.htm//overview (pp. 1–3).

U. S. General Accounting Office. (1999). GAO/GGD-99-170 *Certification requirements, new guidance should encourage transparency in agency decision making.* Retrieved April 25, 2006, from http://www.gao.gov

8

Quality Systems for Testing

Marten Roorda

Cito

Introduction

Objective Assessment

The time will no doubt come when it will be possible to scan the human brain, place all the data in a file, and determine objectively by computer whether a person possesses specific characteristics. In such cases, testing would not be necessary. By that time, as we stroll through the National Test and Assessment Museum, we shall smile piteously at the attributes of a bygone era—a fully equipped testing room, a photo of thousands of Chinese taking a test at separate tables, a stack of stapled sheets of paper with multiple-choice questions.

However, as long as we cannot wire people's heads to be able to download the contents of their brains, we will have to make do with something that is, in effect, artificial—an imitation. Through testing and human assessment, we try to form a good judgment about the person being measured. We make do but there is nothing better. As long as the quality of testing is sufficient, in most cases we can take good decisions. When monitoring the quality of testing, we often make use of international standards. In this section I look at existing quality systems for testing and examine the methods by which tests are judged. I discuss the criteria, the preconditions, and the ethical aspects of international standards. The overview of standards and their assessment will result in a list of recommendations to improve the quality of international standards.

The Aim of Testing

Let us assume that we do not test people simply out of curiosity and that a test serves a specific, predetermined goal. Without presuming to cover everything, I mention several objectives: selection (both based on a norm and on a criterion), diagnosis, self-testing, formative, evaluative, and as a type of survey. On this basis, we hope to take well-founded decisions, such as decisions concerning transfers to higher educational levels, awarding of diplomas or certificates, initiating or adapting treatment, and formulating or readjusting personal curriculum or (national) education policy.

Once the decision to be taken and the aim is clear, one has to ascertain which characteristics of a person qualify for the problem posed by the decision to be taken. Examples of such characteristics are skills, properties, character, knowledge, potential, attitudes, and ambitions. The sum of the qualifying characteristics is the notion or construct that we want to measure. We then operationalize the construct and make it measurable. Instruments will be selected to measure the characteristics. Since, a priori, there is nothing present that can be measured, the measuring instrument should stimulate a response from the person being measured. The behavior that results from this response should be seen as an expression of the characteristic to be measured. The testing situation is somewhat artificial. The situation is not authentic; rather, it is an imitation of reality or it confronts the examinee with hypothetical cases.

Maintaining Quality

Quality is the degree to which something is suitable for use in reaching a specific goal. Whether the quality is sufficient depends on the expectations that we have and the requirements that we establish. The determining factor in assessing the quality of tests is whether it satisfies the objective; that is, whether the quality of the decision that we take on the basis of the test is satisfactory. A test should be good enough for the target set; anything else would be a wasted effort. The success of a test depends on several factors: the reliability of the test as a measuring instrument, the degree to which the test measures what we wish to measure, the efforts and the cost of the test and, finally, the fact of whether the result of the test is desired.

The maintenance of quality takes place in a field of tension. The test publisher and the test user could be inclined to accept lower quality in favor of an economic benefit or improved image.

It is therefore important that those concerned with testing stick to standards, guidelines, or professional codes that not only determine how a test should be constructed and pretested, but also how a test should be used and what will be done with the results. The use of standards, guidelines, and codes is only valuable if they are uniform and transparent. That is why standards are frequently established at an international level.

Codes, Guidelines, Standards, and Assessment Systems

Various Initiatives

There are countless initiatives for standardizing quality criteria for testing, both at national and international levels. The results of these initiatives vary according to the degree that they are peremptory or permissive, as regards the method of approach (process-directed, pro-cedural, or methodological) and as regards motivation and aim. Each attempt to elevate testing to a higher level and make it transparent deserves praise, but the final result is not clearer. Many good intentions have been formulated and much paper printed in an effort to control the construction and use of tests as much as possible. The time has come to map out all those intentions and to evaluate their merits.

Codes of Conduct

Quality criteria could vary from reasonably permissive to quite peremptory. A permissive type is that of professional codes or codes of conduct. This is a cohesive totality of behavioral rules with which specific professional groups are expected to comply in order to exercise their professions in a good, fair, and independent way. An example of this is the Code of Fair Testing Practices in Education (1988, revised 2005), issued by the Joint Committee on Testing Practices. There are also various professional organizations that publish professional codes that do not specifically relate to testing, but issue rulings about them. In none of these instances do members of the profession have to take oaths to a code of honor, as physicians do, for example. Nor, with respect to testing, are they subject to disciplinary rules. Expulsion of members or disbarment to practice their professions is a rare occurrence. I am aware of one case in which a member of the European Test Publishers Group was expelled, but that mainly had to do with copyright infringement. Codes of conduct in the field of testing and assessment are very permissive.

Guidelines

Guidelines that relate specifically to testing criteria go a step further. Although the term could suggest that there are binding regulations or strict instructions, testing guidelines establish the framework within which standards or review systems could be devised. An example comes from the European Federation of Psychologists' Association (EFPA), which published a "review model for the description and evaluation of psychological instruments." Members "may use these as the basis for their own test review procedures. The intention of making this widely available is to encourage the harmonization of review procedures and criteria across Europe" (Bartram, 2002, p. 1). Although the EFPA attaches considerable importance to these guidelines, members are encouraged not to apply them too strictly. "They may be used as they stand or with whatever adaptation is required to conform to local practices relating to test user qualification, test access and related issues" (Bartram, 2002 p. 1). The guidelines should therefore be used as a kind of model standard (as in the case of a model contract), which can be modified according to one's own insights. There is no uniformity or enforcement and there are no disciplinary rules.

The status of the *International Guidelines for Test Use* and the *International Guidelines on Computer-Based and Internet Delivered Testing*, issued by the International Test Commission (ITC), is not clear. These memorandums plot a middle course between a guideline and a professional code (they are not model standards that one can tinker with), but rather seem designed to bring together all current standards and codes to provide greater clarity. "The intention of this document is not to "invent" new guidelines, but to draw together the common threads that run through existing guidelines, codes of practice, standards and other relevant documents, and to create a coherent structure within which they can be understood and used" (International Test Commission, 2000, p. 4). Instead of creating coherence, however, this attempt simply increases the confusion. Here, too, there is no enforcement or sanctions.

Standards

The situation with standards is different, because these are regulations and criteria that professionals are expected to adhere. The best known example of this is the *Standards for Educational and Psychological Testing*, a joint initiative of the American Educational Research Association, American Psychological Association, and the National Council on Measurement in Education. This is an interesting initiative because it goes beyond various professional groups, applies to the entire United States, and has been elaborated in detail. The limitation with standards, however, lies in their power of expression. A test user can claim compliance with these standards, but the standards cannot be imposed on the user. The maximum effect that can be achieved is that end users or test users can demand that test publishers apply these standards. For example, a company that employs assessment in the selection of employees could demand that the psychological consulting firm employ these standards. This would rarely happen, because the customer is more interested in quick results than in what the assessment produces among the job applicants.

Review System

The review system goes one step further. This entails the quality criteria for testing not used by test publishers or test users, but which serve as a guideline for reviewers that determine the quality of a particular test. The most extreme example of a review system is that of the Netherlands Testing Commission of the Netherlands Institute of Psychologists (COTAN), where, besides passive reviewing (the tests are handed in), active reviewing is employed (tests are assessed without reviewers having been invited to do so). Although this does not entail sanctions with an unsatisfactory assessment, the effect could be quite damaging because the review is published. Moreover, in the Netherlands and some other countries, the governments, responsible for good education, require tests to have a satisfactory assessment.

In the Netherlands in some cases diagnostic tests for admittance to remedial education were not approved by the Ministry of Education because of a dissatisfactory review.

Besides formal review systems, there are also informal reviews. These are reviews of tests written by experts, which describe various quality aspects of the tests, accompanied by general, albeit personal, evaluations. An example of such reviews is done at the Buros Centre for Testing. Although its evaluations could be highly personal and subjective, a test could be critiqued that apparently meets all the criteria of a system of standards, which nevertheless is not a good test.

Method of Approach, Motivation, and Aims of Standards

Method of Approach

Systems or standards for quality control of tests could vary considerable with respect to the methods of approach. Each approach has its advantages and disadvantages. Standards, guidelines, and codes can establish criteria for developing tests and for the use of tests. The consideration to emphasize good use of tests is inspired by psychological testing, where a test score in itself is often meaningless, but acquires significance through the interpretation given by psychologists. It should be clear that any test with high construct validity could be botched by means of incorrect test administration conditions, correction procedures, or standard setting. It is therefore advisable to pay attention to both the development and the use of tests.

There are standards or review systems that simply treat tests as instruments. Examples are the EFPA guidelines, the review system of the Netherlands COTAN, and the Buros written reviews. This is good because it elevates the psychometric qualities to a level where they belong. The drawback is that no judgment is given about the application of the instrument. Technically speaking, if there were a test to determine whether someone should receive the death penalty, the test could receive a satisfactory assessment according to the EFPA guideline. Most people, however, would sense that this kind of test would be unethical and reprehensible.

Other systems, such as the guidelines of the ITC, primarily deal with a proper use of testing. The Joint Committee on Testing Practices code is directed both at test developers and test users, but the elaboration is scanty and without obligation. Because standards deal extensively with both test development and test use, they currently provide the most complete overview of quality criteria.

A totally different approach is that of personal certification, whereby the procedures are established for periodic assessment of how well people satisfy their job requirements. The International Organization for Standardization has established an international standard for certification of persons in the ISO 17024 standard. Originating from a European standard, this quality system, which independently recognizes certifying institutes, is growing in global importance. The advantage of this is the independence and the fact that assessment is seen as part of a plan to be able to determine whether people are capable of working in specific professions. It looks beyond the test results.

Accreditation of bodies that certify persons is a good example of external (independent) quality control. In many cases the use of standards is a matter of self-regulation, which takes place internally. Where governments place requirements on testing, such intervention comes from the top down.

Motivation of Standards

Let us not mince words. The world of testing and assessment is an industry formed by businesses. The aim of commercial companies is to maximize profits. There could only be two reasons why test companies control their activities: because they have to or because it produces a benefit. I am not pronouncing judgment on these basic assumptions. They are economic laws from which there is no escape. Still, it is good to recognize the deeper motives of standardizing, because this offers the best starting point when one wishes to assess testing.

It rarely happens that test companies establish controls because they are required to do so. Only in regular education is this fairly common practice. Examples of this are standardized tests and central examinations, including final examinations in secondary education in various countries and entry exams for higher education, particularly in Anglo-Saxon countries. In so far as private enterprises are employed for these types of examinations, there are countless regulations concerning the construction, the standardization, and the administering of examinations. The commissioning agents,

local or national government authorities, usually assume responsibility for this. The role of the test companies is limited to that of a
content provider. It seldom happens that the government establishes
requirements for testing in the private sector because, with the trend
toward privatization, the number of professions that require sworn
officials has declined appreciably. Nevertheless, government controls
are again a growing phenomenon. Governments sometimes want to
take political action in areas where companies do not want self-
regulation. There are examples from the financial sector, but also in
the taxi sector, where the government makes certificates or diplomas
compulsory. In such cases, the government will establish guidelines
or standards that trainers, examining organizations, and certification institutes must satisfy. In this section, however, we do not go
into any further detail about this.

In all cases where the test sector practices self-regulation, it does
so in order to protect the interests of the sector. It is more a negative
motivation than a positive one: it prevents damage to reputations,
covers oneself against damage claims, and enables companies to
maintain fairly stable prices. Dave Bartram (1998), the main author
of the EFPA guidelines, describes the motivation for this as follows:

> The large and growing numbers of test users, with the relative lack of
> provision for good training, combined with the growing international-
> ization of testing, is creating a potentially very serious problem for the
> future. We could see a growth in bad practice leading ultimately to a
> discrediting of psychological and educational testing. (p. 156)

In other cases, there are attempts to employ a more positive perspective by disseminating certain idealism, as in the case of the *Standards*
(1999):

> The proper use of tests can result in wiser decisions about individuals and
> programs than would be the case without their use and also can provide
> a route to broader and more equitable access to education and employ-
> ment. The improper use of tests, however, can cause considerable harm
> to test takers and other parties affected by test-based decisions. The intent
> of the Standards is to promote the sound and ethical use of tests and to
> provide a basis for evaluating the equality of testing practices. (p. 1)

This refers to the interests of all concerned, striving for good and
fair test practice. With the increasing criticism of testing in North
America, particularly standardized tests, this is the least that the
industry can do.

Overview of Standards and Their Aims
ISO 17024

Without attempting to be comprehensive, in this section I examine several of the main standards and quality systems. In selecting these I looked for the systems which are well known and well accepted, including those that have an international status (Code, ITC), the most important system of the United States (*Standards*) and Europe (EFPA), and two examples of review systems, one in the United States (Buros) and one in Europe (COTAN). An overview of the characteristics of the quality systems examined in this chapter is provided in Table 8.1.

In this particular section I start with the ISO 17024 standard, which is the only accepted international system for certification of persons. This was originally a European standard, EN 45013, which the International Organization for Standardization raised to the level of global standard in 2003, without major changes. After having found wide acceptance in Europe, ISO 17024 is currently also advancing in the United States.

The ISO 17024 standard sheet is only available at a very high price—for a paltry 10 pages with general indications. It would be advisable to make this information available on the Internet free of charge, so that there are no barriers to examining it. There is an accrediting body in each country that accredits certification institutes. The standard sheet contains the actual requirements that certification institutes have to meet.

Certification institutes award certificates to persons in order to confirm that they possess the skills for exercising a specific profession, occupation, or task. For professions, occupations, or tasks that one wants to certify, a certification scheme is drawn up that has to be approved by the accrediting body. This scheme describes the necessary competences, methods of demonstrating them, and methods of keeping them up to date. Certain conditions must be met before a certificate is awarded. The plan could include tests, assessments, proof of competences previously attained, endorsement of a professional code, a plan for permanent education (including a system of recognized credits), and any other instrument that offers proof of competence. The duration of validity of personal certificates is limited by definition. In practice, this varies from between 1 and 10 years.

TABLE 8.1   Characteristics of the Quality Systems Examined in This Chapter

System	Educational Testing	Psychological Testing	Region	Technical Criteria	Review System	Can Be Imposed	Active and Anonymous Reviewing
ISO	Yes	Yes	Intern.	No	No	No	No
Standards	Yes	Yes	US	Yes	No	No	No
Code	No	Yes	Intern.	No	No	No	No
ITC	No	Yes	Intern.	No	No	No	No
EFPA	No	Yes	Europe	Yes	Yes	No	No
Buros	Yes	Yes	US	No	Yes	No	No
COTAN	Yes	Yes	Europe	Yes	Yes	Yes	Yes

In some cases an organization other than the certification institute is involved in drawing up the plan. For example, this could be sector or professional organizations. Such an organization must also follow the ISO standards. Motives to control this within a sector could be to promote a professional image, to limit liability, or to achieve a specific cost benefit. An example of the latter is the process industry, which has introduced a safety certificate designed to limit the number of industrial accidents (and their financial consequences). When sectors do not introduce voluntary certification, the system must find acceptance because customers make certificates compulsory for their suppliers.

Many see ISO 17024 as fairly complicated, because it requires a strict separation of roles. Besides accrediting and certifying institutes, there are also examination and training bodies, a board of experts and a board that evaluates the scheme for the certifying institute before it goes to the accrediting body. This entails a considerable amount of conference time and is quite expensive, because commissions are paid along each link of the certification chain. Certain roles may not be combined, such as those of the trainer and certifier, as paragraph 4.2.5 of the standard sheet states: "The certification body shall not offer or provide training, or aid others in the preparation of such services, unless it demonstrates how training is independent of the evaluation and certification of persons to ensure that confidentiality and impartiality are not compromised" (International Organization for Standardization, 2003).

Whereas testing standards that good tests have to meet are often described to the smallest detail, ISO 17024 only gives them cursory treatment:

> The certification body shall evaluate the methods for examination of candidates. Examinations shall be fair, valid and reliable. Appropriate methodology and procedures (such as collecting and maintaining statistical data) shall be defined to reaffirm, at least annually, the fairness, validity, reliability and general performance of each examination and all identified deficiencies corrected. (ISO/IEC 17024:2003, p. 4)

Furthermore, requirements are placed on examiners, which must be familiar with the scheme and everything relating to it, and who must be independent. No requirements, however, are placed on a person's testing expertise. That is odd when seen in the light of the ISO standard, because one might at least expect certification of examiners.

This is dealt with briefly on the test instrument itself. The standard sheet states that a practical analysis must be made at least once a year, which must contain:

> a specification for the construction of the examination(s), where a formal oral or written examination forms part of the evaluation process, including content outline, type(s) of questions being posed, cognitive level(s) of the questions, number of questions for each subject, time length of the examination, method for establishing the acceptance level of the mark, and method(s) for marking. (ISO/IEC 17024:2003, p. 9)

All examinations must meet the specifications, so that they can be uniformly applied and are free from bias.

The advantage of the ISO 17024 system is that it goes beyond the measuring instruments by overseeing all requirements necessary for personal competences. It is also good that certificates have limited durations of validity and that there are requirements established for permanent education. Among the disadvantages is the bureaucratic nature of certification (which also drives the pricing) and the fact that the system only focuses on the procedure and not on the content and yields. Accreditation and certification are not a guarantee of quality. A certification or examination institute can fully comply with the standard and nevertheless administer worthless examinations.

Standards for Educational and Psychological Testing

The aforementioned *Standards for Educational and Psychological Testing* came into being in 1954 when the American Psychological Association published its Technical Recommendations for Psychological Tests and Diagnostic Techniques. The NCME and the AERA joined the APA a year later by publishing a standard together. New versions were issued in 1966, 1974, and 1985, before the current edition appeared in 1999. Because of the long history and extensive detailing of these standards, they have acquired high status with observable influence on other standards and guidelines. Although the *Standards* are for use in the United States, they have had influence on European quality systems, forming the inspiration for EFPA and COTAN guidelines.

The introduction to the Standards explains their objective—or rather, what one hopes to achieve with them: "The purpose of publishing the Standards is to provide criteria for the evaluation of tests, testing practices, and the effects of test use" (Standards, 1999, p. 2). The book actually serves as a kind of checklist so that nothing is forgotten. "Although the evaluation of the appropriateness of a test or testing application should depend heavily on professional judgment, the Standards provide a frame of reference to assure that relevant issues are addressed" (Standards, 1999, p. 2). Although the Standards are not prescribed or imposed on test developers, test publishers or test users, the compilers hope that they will be adopted and encourage others to do the same. In addition, transparency is also a major aim. The intention is not to give psychometric answers to political or social questions, but to provide relevant technical information for use in public debates.

Part I of the *Standards* concerns "Test Construction, Evaluation, and Documentation." The subjects include the validity and reliability of tests, test construction and revision, scales and standards, correction, reporting, and documentation. Each subject is preceded by background information. There is a practical approach: each standard is a concrete, concise guideline with an explanation. For example: "If validity for some common or likely interpretation has not been investigated, or if the interpretation is inconsistent with available evidence, that fact should be made clear and potential users should be cautioned about making unsupported interpretations" (Standards, 1999, p. 18). With this kind of standard, users understand the limitations of a test. It is much like the inserts that accompany medicines. One can discover what the indications and contra-indications are, along with possible side effects. As long as it is transparent for which problems a test provides a solution and for which it is does not, people can decide for themselves whether they want to use a particular test.

Part II treats "Fairness in Testing," with subjects such as justice, the rights of examinees, differences in language mastery, and how to deal with examinees who have disabilities. In part III, there is a more detailed discussion of "Testing Applications." The subjects include the responsibilities of test users and the specific aspects of testing in various areas, such as psychological testing, educational testing, and testing of employees.

The advantages of the *Standards* are high status, large public support, their extensive detailing, and the practical application of the contents. Another seeming benefit is that the *Standards* cross all professions and are not, as is the case with most guidelines, intended only for psychologists. The drawbacks are the noncompulsory nature of the Standards and the absence of a related review system.

The Code

The Code of Fair Testing Practices in Education was introduced in 1988 by the Joint Committee on Testing Practices. The Code underwent revision in 1999, inspired by revisions in the *Standards* that same year. The aim was to make the Code consistent with sections of the *Standards* that relate to education. The question of the added value of the Code was answered as follows:

> The Code is not meant to add new principles over and above those in the standards or to change their meaning. Rather, the Code is intended to represent the spirit of selected portions in the Standards in a way that is relevant and meaningful to developers and users of tests, as well as to test takers and/or their parents or guardians. States, districts, schools, organizations and individual professionals are encouraged to commit themselves to fairness in testing and safeguarding the rights of test takers. The Code is intended to assist in carrying out such commitments. (Joint Committee on Testing Practices, 2005, p. 4)

In short, test experts in education needed something for themselves. The document takes the form of a code of conduct that puts additional pressure on people, in particular to adhere to the standards.

The Code only covers two pages, without any elaborated list of criteria or technical starting points. It concerns professional attitude, however, the good will and intentions of test developers and test users to pursue high quality and the fair use of tests. Part A deals with the development and selection of tests. Part B treats the administering and correction of tests. Part C involves the reporting and interpretation of test results. Part D concerns the information given to examinees. All four parts contain sections for test developers and sections for test users.

The advantages of the code are its easy reference and idealistic nature. Disadvantages are that the code could easily be seen as an uncalled-for expression of political correctness to which one does not necessarily have to adhere; and that the code has little to add to existing standards and guidelines.

International Test Commission

To determine its position, the International Test Commission (ITC) should be described as an association whose member organizations and individual members are involved in psychological testing. The members come from various countries throughout the world.

The aim of the ITC is to assure "uniformity in the quality of tests adapted for use across different cultures and languages" (International Test Commission, 2000, p. 3). This description clearly refers to the problems of translation and adaptation. Following preparatory work by Van de Vijver and Hambleton, in 1995 during a meeting, the ITC decided to expand this aim to include "guidelines on the fair and ethical use of tests, from which standards for training and specifying the competence of test users could be derived" (International Test Commission, 2000, p. 3). The stated long-term objective was to have the guidelines specify the competences that test users have to satisfy. Additionally, these competences should be linked to measurable criteria, on the basis of which a person who wishes to qualify as a test user can demonstrate his or her competences. The competences have nothing to do with the instrument itself, but rather with the choice of instrument, professional and ethical conduct, the rights of examinees, test administration and correction, and writing reports. Because the ITC guidelines are primarily based on test usage, it is clear that they are mainly intended for psychological testing.

It ultimately lasted five years, to 2000, before the ITC guidelines were ready under the title of *International Guidelines for Test Use*. The entire development process betrays the characteristics of a consensus product. Although this long-term view suggests that one is heading for official certification of test users, the 2000 guidelines especially had to avoid erecting too many impediments:

> These guidelines are to be seen as supportive rather than constraining. We need to ensure that the guidelines embody universal key principles of good test use, without attempting to impose uniformity on legitimate differences in function and practice between countries or between areas of application. (International Test Commission, 2000, p. 5)

The guidelines can be used if one wishes, but they can also be used to develop local standards.

The first part of the guidelines revolves around taking responsibility for the ethical use of tests. This extends from having the proper competences as a test user to and including the confidential use of tests and test results. The second part concerns the following of *good practice* with test usage. Among others, this includes the justification of test choice, the use of technically sound tests, attention for fair testing (free from cultural or ethnic bias), arranging equal test administration circumstances, and interpreting and reporting the results.

In 2005 the ITC added a CBT guideline, International Guidelines on Computer-Based and Internet-Delivered Testing. Analogous to the Code of Fair Testing Practices in Education, this document was divided into guidelines for different target groups: test developers, test publishers, and test users. What stands out in comparison to the 2000 guidelines is that an entirely different design was used. The guidelines are elaborated in much greater detail and the nature and scope of the new document are more similar to a standard than the older document, which especially resembles a professional code. The problem is that the huge differences between the two publications make it difficult to discover a single line in the efforts of ITC. It would be better to combine both guidelines in a single document.

The advantages of the ITC guidelines are that they are internationally supported and based on the competences of the test user. This provides common ground for enabling certification of this group. The disadvantages are that the guidelines are hardly peremptory and there appears to be little consistency in the ITC activities.

EFPA

During its conference in 1995, the European Federation of Psychologists' Associations (then still known as EFPPA) set up a task force to develop European policy for tests and testing. This initiative was partly inspired by the free mobility of labor within the European Union and the European directive for supporting this free mobility with a fair selection process. This development ran almost parallel with that of the ITC guidelines. Following a survey among members, a subsequent task force was set up, charged with developing the guidelines.

There is collaboration with the ITC (Bartram, 1998): "This task force is working closely with the EFPPA and the ITC projects, to help ensure that all work progresses in a fashion which will best ensure a successful outcome and some international consensus" (p. 159). That this is a very close working relationship is clear from the fact that the main author of the EFPA guidelines, Dave Bartram, was also the chairperson of the ITC at the time. Although there is absolutely no reason to question the integrity of those concerned, the combination of jobs creates the appearance of overlapping interests. Nevertheless, there is a major difference between the EFPA and the ITC standards. Whereas the ITC mainly focuses on the competences of test users, the EFPA has devised a form for reviewing tests, complete with a set of instructions for reviewers.

The EFPA Review Model for the Description and Evaluation of Psychological Instruments was largely modeled on the then authoritative standards in Europe: the test criteria of the British Psychological Society (BPS) and the Review System for the Quality of Tests of the COTAN, the Netherlands Testing Commission of the Netherlands Institute of Psychologists. Dave Bartram, by the way, was also one of the two authors of the BPS criteria, developed for the UK Employment Services.

The Review Model bears all the characteristics of a score card, geared toward assessing test instruments. Part 1 contains the description of the instrument, including the aim of the test, the nature of the supervision, the item type, the number of items, the response method (paper, computer, oral), the correction method, and the scales used. Attention was also given to computer-generated reports, delivery terms, and costs. Part 2 contains the criteria for evaluating the instrument: the assessment of the materials, the norm or reference groups, validity, and reliability. In part 3 the scores of the evaluation are collected in a single table in order to reach a total judgment about the instrument. Part 4 is intended for recommendations, incorporated in degrees of utility. These vary from low- to high-stake applications.

In the "Notes for Reviewers," an appendix with the review model, assessors receive instructions and regulations for evaluating the instrument. There is a brief description of the Review Model on EFPA's Web site (www.efpa.be) concerning the review procedure. While, for instance, a strict regulation applies for reviews at the Buros Centre for Testing, the EFPA employs the standards rather loosely. "Because countries in Europe differ so much in their resources and facilities for carrying out test reviews, EFPA have not devised a common review procedure." It seems quite reasonable that there should be less extensive review procedures in European countries where fewer resources are available, but it could undermine the entire model if the criteria are always applied differently.

The EFPA does have certain guidelines for the review procedure. For example, reviewing must take place by qualified persons with "a good understanding of psychometric issues." It does not state that these have to be psychometric experts per se—a requirement that Buros insists on. Good governance dictates that reviewing should not be done by people who are associated with the test publisher. Assessment must also be done by at least two reviewers. Ideally, a third reviewer should be involved to compare the two reviews and resolve any discrepancies. If the EFPA review form were to undergo its own procedure, the result might not be so favorable, because the evaluation is a motley mixture of criteria and ratings. There is no clear standard present.

The advantages of the EFPA model are that they are specifically geared toward the quality of the instrument with a review procedure, which is more than merely a guideline or standard. Drawbacks are that the procedure is meager and exceptions may be made per country.

Example of a National Standard: COTAN

Europe, Great Britain, and the Netherlands have the oldest traditions with respect to test standards. While the American Standards originated in 1954, since its founding in 1959 the Netherlands Testing Commission (COTAN) of the Netherlands Institute of Psychologists has been describing tests. The basis for the current system was laid in 1982, with assessment based on five criteria. This system underwent a thorough revision again in 2004. With its Review System for the Quality of Tests, COTAN focuses systematically on the quality of the instrument itself, especially its psychometric aspects. The aim of the system is twofold:

> Test users are informed by means of these assessments about the quality of the instruments. Via the assessments, the test authors receive feedback about the quality of the instruments they develop. The assessment system could serve as a guideline for them in developing tests and writing manuals. (COTAN, n.d.)

Interestingly, for most tests used in education, the Dutch Department of Education makes a favorable assessment by COTAN compulsory.

One endeavors to obtain the highest degree of transparency in the procedure. The assessment system is open to the public. Typically, COTAN conducts policy actively: tests can be submitted, but COTAN also evaluates instruments at its own initiative. A test can be included in an assessment procedure as soon as it becomes available, even if the documentation or technical accountability is still lacking. Further, the high level of confidentiality is striking. It is even the case that the two persons who assess tests remain anonymous. They are designated by the COTAN, which makes sure that they are in no way related to the test publisher, including being competitors. The reviews are published in the Documentation of Tests and Test Research book. The procedure is free of charge to the publisher.

The original five assessment criteria have now been expanded to seven. The judgment accompanying each of these criteria could be unsatisfactory, satisfactory, or good. Some test publishers, as is the case at my own company (CITO), establish an internal requirement that all seven criteria must at least receive satisfactory assessments. An unsatisfactory assessment of a criterion, however, does not mean that a test cannot be used. An unsatisfactory assessment could be given because information is lacking, such as pretest data, which could be added later. Moreover, a test could be used in a part of the scale or in combination with another instrument. When a criterion is assessed as unsatisfactory, it mainly serves as a signal. The intention is to make the test user aware of the value of a test and therefore not to attach conclusions or base decisions on those that do not justify the results.

The review system works as a checklist, whereby a negative assessment with one of the basic questions (has sufficient information been provided) immediately results in an unsatisfactory assessment for the entire criterion. The first criterion evaluates the starting points of the test construction: users must be able to judge whether a test fits the goal for which the test has been chosen. The second criterion is the quality of the test material and the third, the quality of the manual. The fourth criterion concerns the standard setting. If a test is criterion based, the criteria must be clear; and if a test is based on norms, the norm group must have sufficient quality and the ability to generalize must be in order. The fifth criterion concerns the reliability of a test, which is related to taking decisions with the help of the test. The sixth criterion is the construct validity: do the results make it sufficiently plausible that the notion in question will be measured? The seventh and last criterion is criterion validity: are the results satisfactory, given the intended type of decisions that have to be taken with the test?

The advantages of the COTAN review system are the high degree of objectivity and confidentiality, active policy, and psychometric depth. A disadvantage is the approach from psychological testing, although the review system is also used for educational testing.

Example of a Review System: Buros

The assessment of tests by means of written reviews contrasts greatly with precisely described review systems such as EFPA and COTAN. Still, the presence of exact criteria does not have to be a disadvantage. Written reviews make it possible to give a general impression and integral judgment, while review systems that can evaluate tests as satisfactory on technical grounds, despite the fact that they do not make much of an impression when they are used. The oldest, most influential review institution can be found in the United States. It is even older than the Standards.

The Buros Centre for Testing is based on the legacy of Oscar Krisen Buros, who published the first Mental Measurement Yearbook in 1938. The aim of this independent, university-associated institute is *to improve the science and practice of testing* in the interests of test users and the public at large. When the institute was founded in 1994, the fields of research and reviews were expanded from psychological to educational testing, certification, and human resources (such as recruitment and selection).

Although exact criteria are lacking, Buros recommends that reviewers work in specific categories. The first category is the general description of tests, such as the goal and the target group. The second category contains a description of the test-development method, including pretesting and selection of items. The third category deals with technical aspects such as reliability, validity, and standardization: information about norm groups and being able to generalize them to specific target groups. The fourth category provides room for commentary by the reviewer on the strengths and weaknesses of the test. In the fifth and last category, the judgment is summarized in a maximum of six or seven sentences. This is a conclusion, with or without recommendations. A recommendation could even entail stating that another test would be more suitable for the intended goal.

Anyone may sign up at Buros as a reviewer. Buros assigns a reviewer for each assessment, thereby placing high demands on reviewers. They must have demonstrable psychometric expertise and be independent. Evaluations must be fair: compliments given for what is good, but criticism if it is not good. Sometimes a second reviewer is assigned, who writes a separate review. Test publishers receive drafts of the reviews and are invited to comment on them. The idea, however, is not to influence the review. The comments are only intended to correct factual inaccuracies.

The published reviews are available by purchasing the Mental Measurement Yearbook or the electronic version via the Internet. Nowadays, test publishers may cite up to 50 words from a review, if the citation is consistent with the general tenor of the evaluation.

Advantages of the Buros system are the independent, critical position taken and the possibility of giving more general, personal judgments about tests. Disadvantages are that the assessments are not based on standardized criteria and that the procedure is not confidential: the reviewers are known and there are contacts with test publishers.

Preconditions and Criteria for Reviews

Criteria for Reviewing Tests

It is remarkable that, with the large quantity of available standards, guidelines, and codes, only a few are actually geared toward reviewing of tests. Even the rather strict ISO 17024 system only places minor requirements on the quality of testing. It could be coincidental but in general reviewing is never linked to standards in the United States, whereas reviewing of tests in Europe is quite common. The EFPA provides a review model, whereas national institutes such as COTAN in the Netherlands often employ strict criteria and even conduct active policy regarding the reviewing of tests. Another regional difference with respect to criteria is that, in the United States, the emphasis is on fair testing—that is, on test usage—whereas in Europe the emphasis is geared more toward the quality of the instrument. It would seem that in America standards are employed to provide minimum liability coverage for the industry. As long as the test material is political correct and all parties concerned show good will, nothing else is needed. Reviews are seen as too threatening to economic interests, as an infringement of the free-market economy. Fortunately, there is independent reviewing, such as Buros, but these reviews are not open to the public at large; and the end user has to pay a fee to gain access to them.

As extensive as they may be, even review systems such as EFPA and COTAN are not always complete—for example, because they are geared toward psychological testing and not educational or staff testing. Minimum criteria are required to arrive at a proper review. To highlight them it is necessary to distinguish the various roles in the entire testing process. There are test constructors and item makers, test publishers, test users and, finally, the examinees. In many cases the roles of the test constructor and test publisher might be combined, but licenses are also frequently granted and tests are translated or adapted. Sometimes the roles of the test publisher and test user coincide, when the publisher brings a product on the market and then administers it itself to examinees. This entails an additional responsibility.

It is important that the test constructor or developer contributes ideas to the test publisher for making the construct operational and deciding whether a test meets its objective. With item production, the right choice of the right type and number of items must be accounted for and items must be free of bias. There must be criteria for zero measurements, for pretesting, and for being able to generalize the norm group. The test material must be standardized and there have to be criteria for the proper use of scales and standard setting. One must demonstrate that the tests are sufficiently valid and reliable.

It is particularly important that the test publisher documents each test sufficiently and indicates the cases or target groups for which the test is suitable, along with the conclusions that can be drawn from the results. This requires technical accountability. There should be transparent keys or correction rules present.

Specific criteria apply to test users, which start with the proper selection of tests. Tests must be administered under equal and fair conditions, even if this takes place in a test center or via the Internet. There should be transparent standards and, particularly in educational situations, clearly formulated criteria for criterion-based tests. Test correction and reports should satisfy specific conditions. If professional interpretation is necessary, for example, by a psychologist, he or she must also meet these criteria.

Confidentiality of Reviews

When a test review requires sufficient authority, confidentiality is an important point. It is expressed in different ways: the independence of the reviewers must be guaranteed and they should preferably remain anonymous, so that they remain independent. Test materials, items, and answers must be treated confidentially so that the copyrights of the materials will be observed. However, the institute doing the review must also remain independent vis-à-vis the test publisher. Preferably, there should only be businesslike contacts between the two, for example, when asking for materials.

Two reviewers are frequently used so that the assessment does not depend on the individual opinion of one person. The EFPA even recommends using a third reviewer, who weighs the opinions of the first two and offers a final judgment. It is of course extremely important that reviewers have sufficient test expertise and up-to-date knowledge of what goes on in testing and test administration.

In reviewing tests, it would be advisable to provide the reviewers with transparent assessment models or checklists. Unambiguous scoring methods should enable review models to reach final conclusions with clear qualifications—for example, satisfactory or unsatisfactory results (as with COTAN) or results in recommendations, such as showing degrees of utility (e.g., at EFPA).

Finally, it is important that review bodies have or take the opportunity actively to look for tests that they want to assess. In this way, they can do unsolicited reviews of tests, even if they are not submitted to them. Test publishers might otherwise have a tendency to submit only those tests of which they are proud.

Preconditions with Standards

Besides the normal criteria for developing and using tests, there are several preconditions for good-quality testing that are not usually found in standards, codes, or guidelines. These concern such things as translation, standardization, rules, and supervision, or they occur in the stage that follows test administration and reports.

According to Oakland (2004), around half of all tests are developed in other countries. To make them suitable for use in one's own language and culture, translation and sometimes adaptation are necessary. According to Van de Vijver & Hambleton (1996), just translating them is not possible.

> Translating psychological instruments for use in other cultural and linguistic groups is more involved than simply translating text into another language. Various sources of bias can threaten the adequacy

of translations. Distinctions were made in this paper among three types of bias, depending on whether the bias resides in the construct or its characteristic behaviors (construct bias), in the measurement procedure (method bias), or in the separate items (item bias). Simple translation/ back-translation procedures are meaningful only when construct and method bias do not play a role. (p. 97)

COTAN's review system goes even further by demanding that an examination of equivalence provides proof that the same conception is measured in both the translated and original versions.

Because many standards are based on psychological testing, there is little attention for standardization and establishing cutting scores. Psychological testing is not concerned with whether an examinee passes the test. Cizek (1996) noted that one cannot find much about setting standards in the psychometric literature. Nor is there much about this in the Standards: "It is immediately discernable that the standards do not present an integrated, unified perspective on standard setting that would parallel its treatment on other subjects" (p. 13). In education there are many criterion-based instead of norm-based tests. They are compared to previously established criteria, not to group averages. Standards should not only pay attention to standard setting but should also anticipate ways of readjusting the cutting score, if there is a reason to do so. There is a tendency to make readjustments when there are too many unsatisfactory scores. This is an obstacle to proper enforcement of standards.

The conditions for administering tests get too little attention in all standards. Tests should be administered so that all examinees are treated equally and that the possibilities of cheating are kept to a minimum. Examination rules and regulations should be compulsory. These should specify the minimum number of supervisors or examination leaders to be present and the requirements they must satisfy, particularly with respect to impartiality. The sanctions for fraud or cheating should be known. Further, there should be a description of how corrections take place. The corrector should be impartial and, preferably, there should be a second correction. Examinees should have the right to lodge objections against the procedure or the final result and it should be clear whether and where one can lodge appeals against the decision.

The standards should also establish criteria for the follow-up track. In this way, evaluation of test users and examinees about tests could be used to improve test quality. There is no system that incorporates these opinions structurally in versions of tests. The effects of tests or examinations during training or educational tracks are never taken into consideration in the standard.

Finally, criteria are seldom established for tests that become obsolete. It could happen over time that a test no longer achieves the established aim or that the norm shifts with respect to a specific target group. At COTAN, there is a minimum requirement that tests be renormed after fifteen years, for example, by using data return or by taking a new reference group. With personal certification, the certificate always has a limited duration of validity, but no mention is made of how long it is. There are certificates that are only valid for two years, but the VCA safety certificate (in general use in Europe) is valid for ten years.

In short, there are many potential additions to standards and guidelines. No doubt I have forgotten to mention all of them. I therefore remain open to suggestions.

Ethical Aspects of Standards

Political and Local Circumstances

When I take a long, hard look at ideal standards in this text and the review systems of testing, it is easy to forget that the world in which tests take place is by no means an ideal one. Local circumstances could result in restrictions on applying standards. A complexity of local factors determines the framework within which testing takes place, as Oakland (2004) also acknowledges:

> Conditions external to psychology include the nature of a country's social, political, religious, industrial, and economic conditions; attitudes and values towards science, technology, and individual differences; political and financial support for education at all levels as well as for test development and use; prevailing social problems that may be addressed by test use; and one's national language. (p. 158)

If, for example, one country has a strong egalitarian attitude toward education, it will have less interest in testing, because testing measures the differences among people.

Ideally, countries should have real testing cultures where the use of standards could find fertile ground. A real testing culture is nearly always rooted in test tradition. This tradition is reflected in the colonial past of many African and Asian countries. These former British colonies often have a certain test tradition, whereas this is less evident in former French colonies. Northern European countries have a stronger testing tradition than southern European countries. In the Far East, Japan and China have strong testing cultures. The Chinese even have the oldest history with respect to testing. In the Anglo-Saxon countries, where the largest test companies are still situated, one finds a deeply-rooted testing culture.

On the other hand, resistance to testing could grow in those countries with the most extensive series of test batteries, because it is all becoming so overwhelming and people resist the continuing testing of people.

Nor should we forget that in many countries there are simply not enough resources to administer high-quality testing. To start with, there is simply too little money; or the government has other financial priorities. There is, however, often a lack of professionals that could make up the difference. Oakland (2004): "Quality graduate programs in psychometrics and test development are found in few countries, perhaps no more than ten. Therefore the number of graduate programs and thus the number of students prepared in psychometrics are limited, and graduate preparation is not universally available" (p. 167). There are only a handful of test companies that work with psychometrics at a high scientific level. Institutes that are part of or fully supported by government often do not have the resources to invest in in-depth research and product development.

The spread of test expertise often requires support from international organizations such as the World Bank; United Nations Educational, Scientific and Cultural Organization (UNESCO); United Nations Children's Fund (UNICEF); development banks; and the European Union. CITO conducts such projects throughout the world, whereby the transfer of knowledge is often the most important resource in getting assessment programs off the ground. We must always take into account the fact that, in some countries, the bars have to be lowered somewhat when it comes to quality requirements for testing.

Accessibility of Tests

There are often huge differences in countries between social classes, as a result of which good testing is not available to everyone. The price alone could be a barrier for those who cannot afford it. Difference in socioeconomic backgrounds not only creates a danger of bias in the test results, but also could result in certain groups being excluded from tests.

Standards should at least ensure that there is no test and item bias. Bügel and Sanders (1998):

> With test construction, one should make allowances for the fact that the background knowledge of examinees could differ. Tests should reflect the diversity in background knowledge and the experiences of the examinees for which they are intended. The content of context material should not be so specific that specific groups of examinees are totally unfamiliar with it. Nor should an item contain information that is too specific for a particular culture, if a test is intended for examinees from different cultures. (p. 5)

Standards could contain criteria that pertain to the accessibility of tests. If announced price is so high that it has an elitist effect, it would have a negative impact on the assessment. Test users should also be obliged not to apply any improper selection to the admission of examinees. Where examinees live could even have an effect. Those who live in remote locations could find it very difficult to reach a test center. As a rule, test centers are located in more densely populated, more developed parts of a country.

Responsibility and Accountability

Standards do not often provide definite answers as to who is responsible and liable for which aspects of testing. Bartram (2001): "Who is responsible for checking the cultural adaptation of the test? Who chooses the language—the test user, the test taker, or some third party? What norms should be used and who should decide?" (p. 174).

The keywords with standards and professional codes are *responsibility* and *accountability*. Shohamy shows that there are five different schools of thought in the approach to these notions. First, there is the view that it is not possible for test users to make allowances for all possible social consequences. The best one can do is to limit bias within a test. This approach totally rules out responsibility. The second school implies that one is not responsible for the consequences of tests, but that one should point to their intentions, effects, and consequences. The third approach states that test users must accept all the known consequences. This approach has particularly found acceptance with language testers. The fourth view is that test administrators are not only fully responsible for the consequences of their tests, but that they should also expect sanctions if their tests conflict with the standard for proper testing practice. In the worse-case situation, a test could even be prohibited. The fifth school of thought assumes a shared responsibility: "A different view follows the notion of shared responsibility—the responsibilities for good conduct are in the hands of all those who are involved in the testing process" (Shohamy, 2001, p. 147). The advantage of this idea is that none of the parties in the testing process can arbitrarily shift responsibility to someone else.

Conclusions and Lessons Learned

The Future of Standards

Most standards, guidelines, and codes have only been marginally revised lately. There are editorial revisions and the ITC guidelines for computer-based and Web-based testing are the latest developments in this field. It seems as though the industry is now reasonably satisfied with the way it regulates itself. However, in its 2003 Report to the General Assembly by the Standing Committee Action Plan, the EFPA recommended adopting American standards: "APA/AERA/NCME guidelines: Discussion and analysis of a possible European adaptation."

However, several developments should result in revisions of the standards and their stricter application. First, there are the technological advances. Testing is increasingly taking place by computer and even over the Internet. In the future, management of item banks and composition of tests will be quite different. Adaptive testing places different requirements on accountability. Taking security measures will result in administering *high-stakes* testing over the Internet with a remote proctor, or even without a proctor. Secondly, at the request of buyers, publishers are expanding their product package beyond traditional testing. Assessment increasingly takes place on the basis of competences. Other instruments are needed and more human assessment will be used to measure skills, attitudes, and potential in addition to knowledge. This involves providing evidence that someone is capable of performing a given task or profession. In the third instance, testing is constantly coming under pressure—not only by those who are opposed to measuring people or by opponents of standardized tests, but also by the fact that an increasing number of people are challenging the results and consequences of tests in the courts, with a resulting higher degree of liability.

The True Value of Standards

I have often noted that the disadvantages of standards and guidelines are that they are without obligation. Having a review system is really a condition for taking standards seriously. Otherwise there is only a carrot and no big stick. Although there is a review system at EFPA, it will only apply when an associate member translates it into national standards. That only happens incidentally, however, as with the Dutch COTAN. Many guidelines pay little attention to the quality of the testing instrument, even though this is where everything starts. If the instrument has no value, the testing we do is worthless and we could draw the wrong conclusions from it: incorrect diagnoses, incorrect passing or failing of examinees, and hiring or rejecting employees undeservedly. A third disadvantage that recurs with many standards and guidelines is that they are based on psychological testing, while most testing takes place in education or business. Such testing entails requirements other than those of psychological testing, requiring the establishment of broader standards. Finally, many criteria and preconditions are lacking before the standards can cover the entire testing process.

Recommendations

Following this analysis of current standards, my conclusion is that much change is needed before these can be taken seriously. I therefore arrive at the following recommendations.

- Standards and reviews should be available free of charge to everyone and this should be communicated actively to users and end users.
- A review system should always be linked to a standard or quality system.
- Reviewers should always remain anonymous. There should be a minimum of two independent, expert reviewers.
- Active reviewing should be the starting point, including unsolicited assessment of tests.

- Reviewing of tests should always take place outside the sector or industry. Both the materials and the people involved in reviews should be independent of the industry in question.
- When reviewing a test, look at the effects that this test might have on training, education, and performance of those concerned (consequential validity).
- With reviews there should be no contact between the reviewers and the parties concerned, such as test publishers, except when this entails asking for materials.
- Test users, assessors, examiners, and proctors should be certified, so that it can be determined that they possess sufficient competences to perform their tasks. It should also be considered that psychometricians and test constructors need certification, to ensure quality of test development.
- There should also be new criteria for measuring competences, such as human assessment.
- Computer and Web-based testing should be incorporated in the standards. They result in specific criteria.
- Standards should not only be based on psychological testing but also on tests in education and business. More attention should therefore be devoted to standard setting and readjustment of cutting scores.
- In standards, more attention should be given to the way in which tests are corrected, the role of the corrector, the awarding of scores, and the drafting and issuing of reports.
- Translated or adapted tests should always be renormed.
- More criteria should be placed on test administration conditions and the role of proctors. One should also insist that a copy of the examination rules and regulations be present, which includes possibilities for objections and appeals for examinees.
- Tests should be renormed within a specific period; otherwise the tests should no longer be used. The maximum period of validity for a test should be ten years.
- Standards could contain criteria concerning accessibility to tests, such as pricing. Test users should also be obliged not to apply an improper selection procedure for accepting examinees.
- There should be the possibility of sanctions for unsatisfactory reviews or noncompliance with the criteria, such as prohibiting tests or punishing test users or test publishers (disciplinary rules).

Currently in the United States it is probably not feasible to link review systems to standards. The industry would oppose this vehemently. In

Europe, where there are existing test review systems in use, it would seem advisable first to agree on a cross-border standard with a review system, based on the previously mentioned recommendations. A good platform for this might be the Europe Division of the Association of Test Publishers, which was recently founded at my initiative. I would very much like to promote such a European standard for psychological, educational, and personal testing, linked to an active review system.

Acknowledgments

For Carolien, who is my personal standard.

References

American Educational Research Association, American Psychological Association, & National Council on Measurement in Education. (1999). *Standards for educational and psychological testing.* Washington, DC: American Educational Research Association.

Bartram, D. (1998). The need for international guidelines on standards for test use: A review of European and international initiatives. *European Psychologist, 3*(2), 155–163.

Bartram, D. (2001). Guidelines for test users: A review of national and international initiatives. *European Journal of Psychological Assessment, 17*(3), 173–186.

Bartram, D. (2002). *EFPA review model for the description and evaluation of psychological tests, version 3.2b: Review form—Notes for reviewers.* European Federation of Psychologists' Associations, www.efpa.be

Bügel, K., & Sanders, P. F. (1998). *Richtlijnen voor de ontwikkeling van onpartijdige toetsen (Guidelines for Developing Impartial Tests).* Arnhem: Cito.

Cizek, G. J. (1996). Standard-setting guidelines. *Educational Measurement: Issues and Practice, 15*(1), 13–21.

COTAN, Netherlands Testing Commission of the Netherlands Institute of Psychologists. (n.d.). *Beoordelingssysteem voor de Kwaliteit van Tests (Assessment System for the Quality of Tests, August 2004 version). Criteria en Beoordelingsprocedure (Criteria and Assessment Procedure).* www.psynip.nl

Evers, A. (2001). The revised Dutch rating system for test quality. *International Journal of Testing, 1*(2), 155–182.

Facklam, T. (2002, October). Certification of Persons—ISO/IEC DIS 17024, General requirements for bodies operating certifications of persons. *ISO Bulletin*.

Hambleton, R. K., Jaeger, R. M., Plake, B. S., & Mills, C. (2000). Setting performance standards on complex educational assessments. *Applied Psychological Measurement, 24*(4), 355–366.

Hambleton, R. K., & Patsula, L. (1999). Increasing the validity of adapted tests: Myths to be avoided and guidelines for improving test adaptation practices. *Journal of Applied Testing Technology, 1*(1), 1–12.

International Organization for Standardization. (2003). *ISO/IEC 17024: 2003, Conformity assessment—General requirements for bodies operating certification of persons*. Geneva, Switzerland: Author.

International Test Commission. Retrieved from http://www.intestcom. org. International Guidelines for Test Use (Version 2000). International Guideline on Computer-Based and Internet Delivered Testing, (Version 2005).

Joint Committee on Testing Practices, Washington. (2005). Code of Fair Testing Practices in Education (revised). *Educational Measurement: Issues and Practice, 24*(1), 23–26.

Muniz, J. (2003). *Standing Committee on tests and testing, report to the general assembly 2003 in Vienna*. European Federation of Psychologists' Associations. Retrieved from www.efpa.be

Oakland, T. (2004). Use of educational and psychological tests internationally. *Applied Psychology: An International Review, 53*(2), 157–172.

Shohamy, E. (2001). *The power of tests—A critical perspective on the uses of language tests*. London: Longman.

Vijver, F. van de, & R. K. Hambleton. (1996). Translating tests: Some practical guidelines. *European Psychologist 1*(2), 89–99.

Part IV

Design and Planning

9

Designing Error-Free Processes

Rohit Ramaswamy

Service Design Solutions, Inc.

Introduction

In this chapter, tools and methods to create processes that are error free are described. The focus of this chapter is on design, rather than on correction. It is more cost effective and less risky to try, as far as possible, to create processes in which errors do not occur rather than developing extensive inspection procedures to detect and correct errors as they occur. Clearly, there is a trade-off between elimination and correction activities: if it is expensive to eliminate errors and the risks of errors are not so high, there may be circumstances for which detection and correction are better options.

Throughout this book, we use the term *errors* in the broad sense to describe deficiencies in the quality of testing processes. An error or a *defect* corresponds to a condition in which customer or business requirements are not met. In order to decide which errors need to be eliminated from a process, it is necessary to clearly understand the most important requirements for the process, and the risks associated with not meeting these requirements. Many testing organizations have not explicitly prioritized these critical requirements, and therefore no distinction is made between trivial and profound errors. Also, many testing organizations emphasize correction over prevention, and build expensive inspection mechanisms that are still not foolproof in their ability to prevent the most serious errors.

A more effective and efficient approach to error reduction is to focus on using organizational and technological resources to eliminate the possibility of the most serious errors, and then put inspection systems in place to reduce the occurrence of the less serious ones. This requires an understanding of the sources of different types of errors in the test creation, administration and scoring processes, the likelihood of these errors occurring, and the impact of these errors on customers and on the business. This kind of analysis is not often done in the testing industry. In this chapter, a set of analytical and operational techniques used to design error-free processes in manufacturing and service industries and to outline their applicability to the testing industry are described.

Four major steps in designing processes to reduce errors are presented in this chapter. First is a discussion of customer requirements and how to obtain and prioritize them. This is followed by a description of how these requirements, combined with an estimate of the probability of error occurrence, can help evaluate processes to identify the most error-prone process steps, and the areas of greatest risk. This will help determine where to focus error elimination efforts.

Third, a three-step approach for addressing errors in testing processes is presented. This approach attempts to eliminate *opportunities* for making the most serious errors through fundamental changes in the process design; if this is not possible, error-proofing approaches to eliminate the *occurrence* of errors are considered; if this is not possible either, then cost-effective methods to enhance the *detection* of errors need to be designed. Once the process is designed, the fourth step is to manage the process and continually improve it. This chapter ends with a section on conclusions and lessons learned.

Obtaining and Prioritizing Customer Requirements

Ultimately, any product is created to satisfy the needs of its customers. If a critical customer need is not met, then there is little justification for the time and effort spent in designing the product. Tests are no exception. Clearly, if a test does not adequately assess the underlying ability of interest, or if it is not reliable, or if it does not have criterion or construct validity, or if the test booklets or computer presentations of items are of poor quality, then the test will not satisfy the needs of its customers. Similarly, if a test is not delivered on time for administration, or if some answer books are lost, or if the computer system is not stable, or if there are errors and delays in reporting, then the test cannot be considered to be successful. Finally, even a test that meets all the criteria previously mentioned can be unsuccessful for a testing company if it is too expensive or too complex to produce efficiently, or if it requires an unacceptable number of inspections and error checks, or if its security has been compromised.

In summary, a test is unsuccessful if it fails to meet a business or an external customer requirement. Such a test is considered *defective*, or to have *errors*, and we use the terms errors or defects interchangeably in this chapter to refer to situations where the needs of the internal (i.e., business) or the external (i.e., test purchasers or test takers) customers is not adequately incorporated into the design of the test or of its supporting processes.

By definition, error-free tests are therefore those that meet the needs of customers, irrespective of whether these customers are internal or external to the business. When talking about customer needs in this chapter, we need to distinguish between two terms: the *voice of the customer* and *customer requirements*. Even though the voice of the customer and customer requirements sound similar, they mean different things. By customers' *voice*, we refer to the *verbatim comments* made by customers of the product or service. Customer *requirements* refer to the translation of these verbatim requirements into measurable technical requirements. An example to illustrate the difference might be useful. A test taker might say: "I want the diagrams in the test to be clear and easy to read." As a customer of the test, this is the need, expressed in the test taker's voice. For the testing company, however, *clear and easy to read* does not help with the design of the diagram. A more technical definition of these terms is needed. After conversation with the test taker, the test design organization may translate these terms into color scheme, brightness, line thickness, screen resolution, pixel density, and so forth, which are the *customer requirements* for the design. Errors are then defined as attributes that do not meet these requirements. For example, even if the diagram is perfectly correct, incorrect color schemes or inadequate screen resolution or poor line thickness are all errors. The trick to designing error-free tests, therefore, is to clearly understand the voice of the customer and to accurately translate this voice into customer requirements.

Collecting the Voice of the Customer

Where does the voice of the customer come from? It comes from actively engaging with the customers of the testing products and processes, and obtaining information on their needs. It is important to understand what is meant by customers. Depending on the situation, and on the particular test product or process being designed, the customers could be the end users (test takers), test sponsors (state governments, school districts, or professional organizations), regulatory bodies (accrediting bodies, government), test administrators (principals and teachers, training departments of organizations), senior management of the testing company, organizations that are responsible for the downstream processes (e.g., the scoring and reporting organizations are customers of the test development organization), or some combination of the previously mentioned. Given the large number of potential customers and the resources needed for collecting their voice, it is very important to clearly identify the relevant customer set before beginning to collect data on needs.

Collecting the voice of the customer needs to be done through a mixture of reactive and proactive techniques. Reactive techniques refer to the collection of unsolicited information from customers that come through the doors of an organization every day. This information may come from customer complaints, customer feedback after item or test reviews, calls or e-mails from potential customers into sales or customer service, completed feedback or satisfaction forms from test takers or test administrators after a test is complete, and other sources in the course of everyday interaction between the firm and the outside world. The practical challenge associated with collecting reactive voice of the customer is to establish effective *listening posts* to capture the information that comes in and to set up filters to separate the true information from mere complaints. Reactive data can be biased in that we only hear from those who want to communicate with us, but they are also contextually rich because they provide comments in the context of their experience with the company. This makes the feedback immediate and heartfelt and, in many cases, this kind of feedback may have greater resonance than the distant and theoretical feedback often obtained through structured data collection techniques such as surveys or interviews.

This discussion about the reactive voice of the customer is not to imply that testing organizations do not traditionally pay attention to it. Most testing organizations are sensitive to customer feedback, and act quickly to address customer complaints. But in most organizations, the only action that is taken based on reactive data is to solve the problem immediately—there is no systematic effort to collect data, to analyze it for patterns, trends and themes, and to use this information to help design better processes in the future. Our experience is that organizations that do so generate fewer complaints and errors over time, and are more proactively able to align their operations with the needs of the customers.

Because reactive data is biased, it alone is not adequate for making effective design decisions. Reactive data serves a purpose, but it needs to be supplemented with proactive, planned, structured data collection techniques. When people talk about structured data collection, they immediately think about surveys. But surveys are very narrow and limited data collection devices, and only serve to reinforce or prioritize what is already known. More effective proactive data collection techniques fall under the category of *narrative techniques*. These are methods to elicit different types of information depending on the context. Interviews are one form of narrative technique, and are used to elicit broad themes and issues that can be elaborated on by using other techniques, such as focused elaboration, most significant change (Davies, 1998), appreciative inquiry (Cooperrider & Whitney, 2005), or contextual inquiry (Beyer & Holtzblatt, 1998). These methods work best if they incorporate both observation and interrogation, that is if they elicit information by asking contextual questions based on observation of customers using the test or experiencing the processes. For example, focused elaboration asks specific questions to clarify one aspect of customers' needs expressed broadly in an interview. Significant change methods look at a customers' past experience with a product or process, and seek to identify points in the past where significant positive or negative changes in perceptions or attitudes took place. Appreciative inquiry focuses on positive aspects of future change. Contextual inquiry, like focused elaboration, tries to get deep insights into the voice of the customer by asking carefully structured questions based on observations. Table 9.1 gives some examples of questions that are relevant to understanding needs in the testing industry using each of the narrative techniques described previously.

Analyzing and Prioritizing the Voice of the Customer

Once the voice of the customer has been obtained through one or more of the methods in Table 9.1, several techniques exist to extract

TABLE 9.1   Sample Questions for Obtaining the Voice of the Customer

Proactive Technique	Sample Question
Free Form Interview	What are the most important characteristics of a vendor's test delivery process that you consider before making a decision to select a vendor to administer a state's test?
Focused Elaboration	What do you mean when you say that the diagram on your test screen should be "easy to read?"
Most Significant Moments	Having experienced the long and difficult process of transitioning to your new test vendor, what would you characterize as the most important, good and bad moments of the process?
Appreciative Inquiry	How do you see the future of our partnership in the next 2 years?
Contextual Inquiry	Why do you ship the blank test forms separately from the graded ones when you send them back to be scored?

the needs from the interview transcripts, and to reword, combine, and categorize them. These techniques, such as the *Kano model,* the *affinity diagram,* or the *Voice of the Customer Table* are outside the scope of this chapter. Many books on quality improvement or Six Sigma (see, for example, Ramaswamy [1996] or Pande, Neuman, & Cavanagh [2000]) explain these techniques.

The final step in voice-of-the-customer analysis is to prioritize needs. Not all needs are created equal, and in a world of limited resources, it is important to focus error-reduction efforts in the areas where they are likely to produce the greatest returns. There are many ways of prioritizing customer needs depending on the level of precision needed and the time and resources available. Sometimes it is enough to estimate the importance of a need by tallying the number of times the topic comes up in interviews; at other times, greater precision is required, and a structured questionnaire may be sent out to rate the importance of various needs. In some cases, a panel of experts is used to rate or rank the needs based on their experience. We do not present the different methods for prioritizing needs in this chapter; suffice it to say that it is important to understand the most important needs so that the organization can focus on them.

Generating Customer Requirements

As mentioned earlier in this chapter, by *customer requirements* we mean the quantified, technical description of how the testing processes should perform. These requirements need to be generated from the verbatim customer needs obtained from the voice of the customer through a process of translation. There is no magic formula for performing this translation; the testing organization must review the voice of the customer data, and through a process of discussion must come up with different measures by which to evaluate how well these needs are being met.

The first step in creating customer requirements is to link customer needs back to processes. In testing organizations, as in most other organizations, all work gets done through processes. Therefore, the articulation of what customers want from the organization, whether expressed in the customers' voice or in the form of technical requirements, has to refer to one or more organizational processes. For example, a customer need for accurate or timely reports is associated with the process for creating and/or sending reports. In many organizations, process maps exist for the key processes before customer requirements are generated. If the processes do not exist, or are not documented, the organization may first undertake a high level *process description* exercise that identifies all the processes through which work gets done in the organization. This process description does not need to document the processes in any level of detail; the description could just be a list as shown in Table 9.2.

TABLE 9.2 Examples of Core and Enabling Processes for a Testing Organization

Process Name	Process Type
Author Management	Enabling
Item Development	Core
Test Creation	Core
Test Administration	Core
Test Scoring	Core
Report Generation and Distribution	Core
Item Bank Administration	Enabling
Test Security Administration	Enabling
Sales	Enabling
Employee Recruiting and Support	Enabling

TABLE 9.3   Examples of Customer Needs and Associated Processes

Process	Customer Need (Voice of the Customer)
Item Development	Sufficient items by content area need to be available every time a form needs to be created/assembled.
Item Bank Administration	The usage and psychometric statistics should be available for each item in the item bank.
Item Bank Administration	Only current and relevant items should be available in the item bank.
Sales	It is important to add new clients and new programs every year.
HR	It is important to identify sufficient content experts for writing items.
Test Administration	Answer sheets should be returned for scoring in a timely manner.
Test Scoring	It is very important to make sure that score keys are correct and that scores are accurately computed.

It can therefore be accomplished in a relatively short time. Typically, the senior management team is assembled for a day's workshop to generate this list of processes. The processes may be categorized as *core*, or *enabling* processes, where the core processes are those that directly deal with the creation, administration, and scoring of tests, and the enabling processes are those such as sales, marketing, human resources, or finance, which support the core activities of the organization. Chapter 1 and chapter 13 of this book describe core and enabling processes and the documentation of these processes in greater detail. Once the list of processes has been defined, the customer needs obtained from the voice of the customer can be mapped to the appropriate processes. Some sample needs and the associated processes are shown in Table 9.3.

To translate the customer needs shown in Table 9.3 to customer requirements, appropriate measures are needed. For some needs, the measures are quite obvious and directly translatable form the verbatim need statements; in other cases, they may not be so clear. A good practice is to define *characteristics* for each need, and then to translate the characteristics into measures. Characteristics are qualitative attributes such as *timeliness*, or *accuracy*, or *security*, which describe the core quality requirement behind each need statement. Characteristics are not only useful for generating measures, but also allow us to talk in a technical way about the key aspects of quality that our

TABLE 9.4   Example of Characteristics and Measures

Process	Customer Need (Voice of the Customer)	Characteristic	Measure
Item Development	Sufficient items by content area need to be available every time a form needs to be created/assembled	Item availability	% items that need to be newly created for a new form of a test
Item Bank Administration	The usage and psychometric statistics should be available for each item in the item bank	Data completeness	% items with missing statistics
Item Bank Administration	Only current and relevant items should be available in the item bank	Item currency	% obsolete items in item bank
Sales	It is important to have new programs and new customers every year	Sales growth	% change in new clients from previous year; % change in programs from previous year
HR	It is important to identify sufficient content experts for writing items	Availability of experts	Number of content areas for which experts are not available
Test Administration	Answer sheets should be returned for scoring in a timely manner	Return timeliness	% answer sheets not returned by due date

processes must have. They represent the bridge between customer verbatim statements and precise quantitative measures. Table 9.4 shows the customer need statements in Table 9.3 converted into characteristics and measures; these form a sample set of customer requirements for some of the core and enabling testing processes. As the table shows, the characteristics are technical descriptors of the attributes requested in the customer need statements. Also notice that the measures have been expressed in *negative* terms, that is, as the percentage of units not meeting the requested standard. This is good practice because it draws attention to the defect rate, which is what we are seeking to reduce or eliminate.

Table 9.4 also illustrates some examples of the broad range of characteristics that contribute to test quality. None of the characteristics in Table 9.4 explicitly deal with making mistakes in test items or with errors in scoring or reporting processes. Testing organizations that begin a quality improvement program might choose to focus on these kinds of errors first because they have the greatest impact on the core business of the organization and directly affect the experience of the external customers; however, inefficiencies in the item bank administration, or in the review process, also affect the overall effectiveness of the organization and must not be ignored.

As Table 9.4 shows, there could be a large number of customer requirements at the enterprise level that reflect the quality needs of internal and external customers. No organization has the time and resources to focus on all requirements simultaneously, and therefore it is necessary to prioritize the requirements so that we can identify the top few. Because the requirements are generated from the customer needs, and the needs are prioritized, the most important requirements are those that are associated with the most important needs.

The final output of the customer requirements analysis is the *customer requirements table* shown in Table 9.5. This table summarizes, by process, all the key customer requirements, and their priorities. The rating scale shown in Table 9.5 is typical, with 1 being the least

TABLE 9.5   Example of Customer Requirements Table

Process	Requirement	Average Priority Rating
Item Development	% items that need to be newly created for a test	3.9
Item Bank Administration	% items with missing statistics	3.8
Item Bank Administration	% obsolete items	3.6
Sales	% change in new customers % change in new tests	4.2
Test Creation	% customer reviews that need to be repeated or conducted multiple times	4.5
Test Administration	% forms not returned by due date	4.8

important and 5 being the most important requirement. The individual ratings of a number of customers are averaged to create the scores shown in the third column. Other rating or ranking schemes can also be used.

What does Table 9.5 tell us? Table 9.5 shows that return timeliness, associated with the test administration process, is the requirement that customers consider most important, followed by the completeness of customer reviews and the effectiveness of the sales process. It stands to reason that if resources are limited, they should be assigned to improve these characteristics before focusing on any others.

Another important point is apparent from Table 9.5. In discussing the voice of the customer or customer requirements in this section, we assumed that there is only one customer who is concerned about the outcomes of the organization's process. In reality, there are many customers, or segments of customers. In Table 9.5, the customers who may be most concerned with having test reports returned on time are the test takers. The customers who may be most concerned about not having enough reviews of the test might be the test development organization. And the customers who may want increased sales may be the senior management of the organization. It is therefore important to keep in mind that customer requirements may represent the needs of multiple segments, and that they might not all be compatible. It may be necessary to prioritize further by first focusing on some segments. For example, test takers or test purchaser segments may be deemed to be more important than internal customers.

The importance of the customer requirement is only one variable that influences the choice of where to focus improvement efforts. Another key variable is the probability of an error occurring on an important customer requirement. For example, the timeliness of returns of test forms for scoring is the most important customer requirement, and situations in which returns are not timely are errors in the process. But if the probability of the occurrence of this error is very small, then it would make sense for the organization's quality improvement efforts to focus on other errors that may not be as important but that occur more often. The selection of which errors to focus on therefore depends on a combination of which customer segments are most critical, which requirements are most critical, and how unlikely it is that these requirements can be met by the current process. These interactions are described in more detail in the following section.

Identifying Areas of Risk

The prioritized customer requirements in Table 9.5 identify the processes that are most important to a customer or to a customer segment. The measures associated with the customer requirements pertain to the *output* of a process; that is, they measure the quality of the process required by the customer. But errors take place *within* a process, and therefore to identify areas of risk, it is important to determine how errors in process steps contribute to the overall quality of the output. By identifying the process steps where errors are most likely to occur, we can determine areas of greatest risk in the process and where our error reduction efforts need to be concentrated.

In order to identify the areas of risk, the following steps are used:

- Map the lower level steps for each relevant process to the required level of detail.
- For each step, determine the kind of errors that can take place and the likelihood of these errors occurring.
- Multiply the importance of the customer requirement associated with the error by the probability of occurrence to get a composite risk score.
- Check if any controls are currently in place to mitigate the errors with the highest risk.
- Estimate whether the error is easily detectable even if there is no control in place.
- Decide whether an error mitigation strategy is required for the error.

Mapping the Lower Level Steps to the Required Level of Detail

In the chapter on process management in this book (chapter 13), Ramaswamy describes different levels of process maps. In that chapter, a *Level 2* map is a process flow diagram that describes each process in 5 to 7 steps, and a *Level 3* map takes each of the Level 2 steps down to the next level of detail. There is no magic formula that describes the appropriate level of detail to which we need to map the process for identifying risk areas, but the rule of thumb is to go down to Level 2 or Level 3 or somewhere in between. The actual level of detail depends on what aspect of quality is specified by the customer requirements.

For example, in Table 9.5, the most important customer requirement is associated with the test administration process. The Level 2 process map for the process is shown in Figure 9.1.

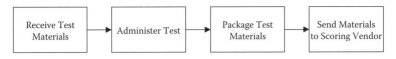

Figure 9.1 Level 2 process map for the test administration process.

In Figure 9.1, the Level 2 map encompasses all test administration activities, from receiving the test form, to administering the test, to sending the test materials back to the scoring vendor. But from Table 9.5, the customer requirement only pertains to the last activity, that is, the sending of answer sheets, because the customers are only concerned that the completed answer sheets are sent back in a timely manner. The process map shown in Figure 9.1 does not provide a detailed enough description of the last step. In this case, a Level 3 process map for the last step of Figure 9.1, shown in Figure 9.2 later, is the appropriate level of detail. This may appear a little artificial, but it is important to note that this is only an example. There will be diverse requirements affecting many different process steps in a real testing application.

Determining the Kind of Errors and Their Occurrence Likelihood for Each Step

What are the kinds of errors that can be made in sending forms back for scoring? Clearly, every step of the process shown in Figure 9.2 can have multiple errors. Once again, the level of detail becomes important. Errors should be identified at a level of detail that allows actions to be taken to mitigate these errors. Here are some errors that may occur in the steps of the process shown in Figure 9.2.

- Failure to schedule shipping
- Failure to deliver or failure of shipping company to pick up
- Delays in the shipping process
- Failure to log receipt of forms
- Materials damaged in transit

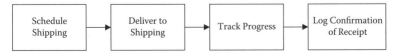

Figure 9.2 Level 3 map of the "Send Materials" process step.

TABLE 9.6   Classifying Errors by Probability of Occurrence

Probability Rating	Frequency of Occurrence
5	Error is likely to occur several times a month (5 or greater).
4	Error is likely to occur several times a year (5 or greater).
3	Error is likely to occur a few times in a year (fewer than 5).
2	Error is likely to occur a few times in 1 to 5 years (fewer than 5).
1	Error is likely to occur a few times in 1 to 20 years (fewer than 5).

How often do these errors occur? A probability rating scale, calibrated in advance, is used to evaluate the likelihood of each error occurring. A typical scale that is applicable in many situations is shown in Table 9.6. It is good practice to recalibrate this table for each application to ensure that it reflects the errors that are characteristics of the particular process under analysis, and that the ratings can truly reflect the frequencies with which these errors occur.

Obtaining a Composite Risk Score

This is the key step in identifying areas of risk. For each error, multiply the probability rating by the importance of the associated customer requirement (obtained from Table 9.5), as shown in Table 9.7.

Table 9.7 shows that the two errors that have the greatest risk are associated with shipping delays followed by failure of the shipper to pick up the shipment. In this example, all the errors are associated with a single process and with a single requirement. In a real analysis, of course, the errors will be scattered across processes and requirements, and the risk levels have to be compared across these multiple dimensions. The composite risk number provides general guidelines for prioritizing areas of risk, but some other considerations may also be important. Some of these considerations are:

- Whether the error, however rare, will result in catastrophic failure of one or more processes with consequences on reputation.
- Whether the error needs to be mitigated as a legal or governmental requirement.
- Whether correction of the error, if it occurs, will take significant time, cost, or resources to fix.

TABLE 9.7 Composite Risk Score for Errors in Sending Tests for Scoring

Error	Associated Process	Associated Requirement	Requirement Importance	Probability Rating	Composite Risk Score
Failure to schedule shipping	Test Administration	Return Timeliness	4.8	2	9.6
Failure to deliver or pick up	Test Administration	Return Timeliness	4.8	3	14.4
Shipping delay	Test Administration	Return Timeliness	4.8	4	19.2
Failure in logging receipt	Test Administration	Return Timeliness	4.8	2	9.6

Other criteria might be important depending on the types of errors and processes involved. Prioritization criteria must always be decided upon in advance before the risk analysis is performed.

Checking Whether Controls or Detection Methods Exist

For the most important errors identified through the risk analysis, it is useful to see if any controls currently exist. Typically, controls use after-the-fact inspection to determine if the task has been correctly performed. The ultimate objective is to attempt to eliminate high-risk errors, or even to eliminate the opportunity for them to occur, so inspection-based controls may not be the ultimate desired solution. But to know they exist provides us a degree of comfort that something is currently being done, and so errors for which controls exist may not be first on the list for error mitigation. For example, a control for late or missed pick-up of test papers by a shipping company may be to remind them if the estimated pick-up time has passed. A mitigation strategy may be to send a reminder before the pick-up time, or to eliminate the need for manual shipping by using computer-delivered tests.

If no controls are present, we need to determine whether it is at least possible to detect errors when they occur. Detection is clearly inferior to error elimination or control, but it is one step better than not being able to tell when an error has been made. For example, there may be no control over delays caused by a shipping company, but tracking the shipment will enable us to detect delays. In this case, therefore the delay is detectable.

In summary, the risk analysis, combined with the determination of whether controls and detection capabilities exist, provides a plan for which errors need to be tackled first. How to go about doing this is the topic discussed in the following section.

Dealing with Errors

As mentioned earlier in this chapter, a three-step approach is used to deal with errors. These steps are: reduce the *opportunities* for errors; control the *occurrence* of errors; or if neither of these approaches is feasible then look for ways to *detect* errors.

1. *Reducing opportunities for error:* This is typically done by redesigning the process so that the steps where the critical errors take place no longer exist or are fundamentally altered. Examples of process redesign to reduce error opportunities are:

   a. Automating steps that are performed manually.

   b. Combining steps so that errors in hand-offs between steps are eliminated.

   c. Reducing decision points by streamlining processes so that mistakes in decision making do not occur.

   d. If there is enough volume, creating separate processes for different products to minimize having multiple paths in a single process (e.g., having a separate process for creating paper-and-pencil and computer-based tests).

   e. Reducing the number of steps overall in *any* process, through simplification.

2. *Reducing occurrences of error:* By following some of the process design principles mentioned earlier, opportunities for errors to take place can be eliminated. But even in perfectly designed processes, variability in execution can lead to errors. Execution errors in processes can be reduced in two ways: (a) by designing safeguards that prevent errors from taking place; (b) by implementing mechanisms that detect and correct errors as soon as they are made so that these are not propagated further in the process. Examples are:

   a. Standardizing processes and work activities and clearly documenting how work flows through the process.

   b. Providing checklists, templates, and other job aids that facilitate the doing of work in a consistent and logical manner.

   c. Color coding exceptions, expedites, and other variations to the normal process, so that these are clearly recognizable as they are being processed.

   d. Incorporating simple auto-check procedures immediately after the task has been completed so that errors can be locally corrected.

   e. Implementing auto-detect and auto-correct technology using logic filters wherever possible.

   f. Introducing just-in-time flow techniques so that work flow is balanced through the process.

   g. Establishing the optimal work speed for systematic on-time task completion and staffing the processes to support this work speed.

   h. Using "forcing functions" so that certain tasks cannot be started until prior tasks are correctly completed (e.g., correct completion of a test review step will result in the issuance of a password that needs to be entered into the system to move to the next process step).

i.   Using auto population of standard information where possible.

3. *Improving error detection:* For errors that are especially high risk, or where the measures described earlier do not completely eliminate the occurrence of errors, it is important to ensure that the errors can be detected as soon as possible, and especially before they become visible to the customer. The general principle is to detect errors in as few steps as possible after they occur. Some of the autocorrect mechanisms mentioned previously (for example, d. and e.) are examples of instantaneous detection. Inspection (often called quality control) is the most common method of error detection, and is already extensively employed by testing companies. Unless carefully planned, however, inspection can be a wasteful activity. It is most efficient to have no more than one or two major inspection points in a process. These inspection points are often critical points in a process, where a major change in activity, or a change of owner or a process hand-off occurs. At these points, it is important to establish a clearly defined process for inspection, with the associated tools and templates so that a thorough identification of all errors can be efficiently accomplished. Depending on the particular nature of the process, it may also be possible to design other mechanisms for error detection. A small, relatively hard-to-detect error can be engineered to trigger a larger, more obvious, and immediately detectable error in a downstream process step. For example, a wrongly entered field in a data form that is used in a later computation can be designed to produce impossible results that are flagged by an exception checking program. If there are logical links between different pieces of data in a database, then an error in one piece of data may result in the disruption of this logic; testing to make sure that all logical links are intact can provide a method for detecting errors. The actual methods may vary from process to process and from testing organization to testing organization, but the fundamental idea of detecting errors as early as possible is universally applicable.

Process Management and Continuous Error Prevention

This chapter has focused on design methods for eliminating the opportunities for errors and for reducing the occurrence of errors in the various processes through which work is performed in testing organizations. Design, however, is only one part of the picture. As discussed in more detail in the chapter on process management in this book (chapter 13), even the best designed processes may fail to perform effectively over time if they are not monitored and managed. Over time, as staff change or as current staff become less vigilant, adherence to standards and to operating practices that result in reduced errors may not be as strong as when the process was first designed. Even if errors continue to be detected and corrected before they reach the customer, drifts in process performance might lead to greater risks and inefficiencies. Therefore, it is important to set up a process management infrastructure to collect data on the errors that occur every week or every month, to compare these errors with those in previous time periods, to make sure that there is no deterioration in process performance, and to continue to look for ways in which errors can be prevented or reduced. This improvement approach supported by monitoring and analysis will ensure that the processes consistently perform as designed. Moreover, the philosophy of always doing better and making sure that the customer needs are always kept in mind will continue to enhance the quality and the reputation of the testing organization over time.

Conclusions and Lessons Learned

The following are key points of this chapter:

- It is more cost effective and less risky to try, as far as possible, to create processes in which errors do not occur rather than developing extensive inspection procedures to detect and correct errors as they occur.
- An *error* or a *defect* corresponds to conditions in which customer or business requirements are not met.
- Processes must be designed based on the *voice of the customer* translated into performance requirements.
- A systematic procedure for identifying areas of risk and vulnerability in processes helps to prioritize areas of focus for error prevention.
- The performance of even the best designed processes can deteriorate over time; it is important to make sure that the processes are monitored, managed, and continually improved to sustain the gains.

References

Beyer, H., & Holtzblatt, K. (1998). *Contextual design: Defining customer-centered systems.* San Francisco: Morgan Kaufmann Publishers.

Cooperrider, D. L., & Whitney, D. (2005). *Appreciative inquiry: A positive revolution in change.* Berrett-Koehler Publishers.

Davies, R. J. (1998). *Order and diversity: Representing and assisting organizational learning in non-government aid organizations.* PhD Thesis. University of Wales—Swansea.

Pande, P. S., Neuman, R. P., & Cavanagh, R. R. (2000). *The Six Sigma way: How GE, Motorola and other top companies are honing their performance.* New York: McGraw-Hill.

Ramaswamy, R. (1996). *Design and management of service processes.* Upper Saddle River, NJ: Prentice Hall.

10

Transitioning between Testing Contractors

Charles A. DePascale
National Center for the Improvement
of Educational Assessment, Inc.

Introduction

Transitions between testing contractors are inevitable in K through 12 statewide student assessment. State assessment contracts are awarded for a fixed and limited number of years on the basis of a competitive bid process. Even when a state is satisfied with its current contractor and has no desire or plans to make a change, a low bid by a reputable testing contractor can result in an unwanted or unexpected transition. Of course, there are also cases where the relationship between a state and its contractor grows stale and both agree that it is time for a change and, unfortunately, cases where such a level of dissatisfaction is reached that the contractor is fired.

Whatever the reason for the change in contractors, the assessment program must go on. In this era of mandated annual assessment and high-stakes accountability, there is little time to accomplish all of the large and small tasks related to a change in contractors (e.g., negotiating contracts, establishing a project plan, ending old working relationships, developing new working relationships, updating address books and contact information) while simultaneously developing, administering, and reporting results from the annual state assessment. All of these tasks must be accomplished, however, and accomplished on time and flawlessly when annual state report cards must be produced, adequate yearly progress must be determined, and, now in the majority of states, decisions must be made regarding student graduation or promotion.

The focus of this chapter is on how states and contractors can successfully manage the transition between old and new assessment contractors. In the first section of the chapter, two critical changes in state assessment that have had a tremendous impact on transitions between testing contractors are addressed: the need to measure progress and the shift in the roles of state and contractor in state assessment programs. In the second section, keys for managing a successful transition are discussed and a case study is presented to illustrate lessons that an individual state applied in planning for and implementing a second transition between contractors. The chapter concludes with a discussion of steps that can be taken by the industry to increase the likelihood of successful transitions.

The Transition in Transitions

Changing Measures versus Measuring Change

It was not long ago that transitions in state assessment programs were mundane, particularly from a technical perspective. Through the 1990s and into the initial years of the 21st century, whether a state's assessment consisted of a commercial, norm-referenced standardized test (NRT) or a custom assessment, when a major change in the assessment program occurred there was a common response—start over again. In the case of state assessments, starting over again meant a break in results. There was no direct link between the results on the new assessment and results on the old assessment. There were a variety of approaches to starting over again. Some approaches made the change quite obvious and others were designed to make it appear that no change had occurred.

Among states administering commercial NRT, there were three common occurrences that resulted in change: a current test was renormed, a new version of the same NRT was introduced, and a different NRT was adopted. The most common metrics for reporting NRT results are percentile ranks, normal curve equivalents, or grade equivalents. When a change occurred, states shifted from the old norms to the new norms and reported the same statistics. Nobody was surprised by a shift in results (either up or down) from the previous year. A conversion table might be produced in the first year after a change to show the relationship between new and old results, but results were reported in terms of the new norms and no attempt was made to create an ongoing link between old and new results.

Among states with a custom state assessment, change most often meant the introduction of a new assessment program or dramatic changes to the format of the current program (e.g., the introduction of constructed-response items to a multiple-choice test, changes in state content standards). Common metrics for reporting results from custom state assessments are scaled scores and, more recently, distributions of students across achievement levels. When a new state assessment program was introduced, a new acronym was created, a new reporting scale was developed, and new achievement standards were set—each usually different enough from its predecessor to avoid confusion. On the other hand, when a significant change was made to an existing program, a state might wish to retain the look-and-feel of the old reporting scale even when a direct link between the tests was not the optimal measurement approach. Under those circumstances, a new reporting scale was developed that maintained the same statistical properties as the old scale and some method of equipercentile equating was used to generate state results consistent with the previous year.

In that earlier period of large-scale state assessment, lack of annual growth at the state level was a psychometrician's best friend. Scores now regarded with alarm as stagnant were then looked upon as stable and steady. Because little or no change was expected in scores at the state level from one year to the next, it was a simple task to produce a new score scale that looked identical to the existing scale and resulted in the same distribution of students across achievement levels. There were changes in results at the district and school levels, of course, but educators and the public were cautioned to interpret year-to-year changes in results with caution and to allow time for schools to adapt to the new format or new standards.

As an accountability wave began to spread across the country in the 1990s, however, a direct link between annual assessment results became more critical (Rothman, 1995). A few states (e.g., Kentucky, Louisiana, Massachusetts) began to build school accountability systems that required improvement in assessment results over time. The 2001 reauthorization of the Elementary and Secondary Education Act (ESEA)—commonly known as No Child Left Behind (NCLB)—effectively eliminated the option to just start over again for all states (PL-107-110). To measure adequate yearly progress it was now necessary to measure progress, improvement, or growth—defined in terms of annual changes in assessment results. The basic assumption that state assessment results would remain stable from one year to the next was replaced with the assumption that those assessment results should and would show improvement each year—in most cases, extraordinary improvement.

Vendors versus Contractors

There has also been a shift since the early 1990s in the relationship between the state and testing contractor that has increased the complexity of transitions between testing contractors. The traditional relationship between the testing company and state can be characterized as that of vendor and customer. The state and testing company negotiated a price for the use of the testing company's test. In many cases, the current relationship between the testing company and the state is better described as contractor and client. The testing company is hired by the state to provide expertise and services needed to help build and administer the state's test in the same way that a general contractor is hired to provide the expertise and services needed to build a customer's new house.

Prior to 1990, not all states administered a state assessment and those that did relied primarily on multiple-choice, norm-referenced tests (Olson, 1999). In most cases, the state purchased an off-the-shelf NRT and scoring services directly from a testing company. The test was designed, developed, and owned by the testing company. Following the purchase and scheduling of the test, it would not be unusual for the state to have little direct contact with the testing company or involvement in the operation of the testing program. The testing company would deal directly with local districts or schools to arrange for the shipping and receiving of test materials and reports. Following testing, the state received a report of local district or school results as well as a state summary from the testing company.

Beginning in the late 1980s, several factors related to the broad concept of education reform contributed to a shift in the traditional vendor–customer role between the testing companies and states. Some factors impacted the nature and format of the tests themselves. A backlash against multiple-choice tests in favor of more authentic forms of assessment resulted in changes in the design of tests.

Criticism of norm-referenced reporting, exemplified by the coining of the *Lake Wobegon Effect*, contributed to a shift toward the reporting of test results in terms of achievement levels rather than percentile ranks and grade equivalents (Miller, 1997). Other factors impacted the role of the state in local education. State-level education reform efforts responding to reports such as *A Nation at Risk* and school finance lawsuits changed the role of the state and the role of state assessment (Hurst, Tan, Meek, & Sellars, 2004).

Federal legislation also played a significant role in changes to both the state assessments and the role of the state. The 1994 reauthorization of ESEA known as the Improving America's Schools Act of 1994 (IASA) redefined state assessment by requiring states develop or adopt "a set of high-quality, yearly student assessments" for use as the primary means of determining the adequate yearly progress of schools served by Title I (PL 103-382). Additionally, IASA included the following requirements for states regarding state assessments and the content and performance standards that the assessments were designed to measure:

(a) develop or adopt challenging content standards and challenging student performance standards for all students which shall include the same knowledge, skills, and levels of performance expected of all children;

(b) ensure that the state's challenging performance standards are aligned with the content standards and describe two levels of high performance (Proficient and Advanced) and a third level of performance (Partially Proficient) to provide complete information about the progress of lower achieving children;

(c) ensure that the state assessments are aligned with the state content and performance standards; provide coherent information about the attainment of those standards; involve multiple up-to-date measures of student performance, including measures that assess higher order thinking skills and understanding; provide for the participation of all students; and provide for the reasonable adaptations and accommodations for students with diverse learning needs. (Title I, Part A, Sect. 1111.)

The 1997 reauthorization of the Individuals with Disabilities Education Act (IDEA) further refined the nature of state assessment with the requirement that states ensure that all students are able to participate in the state assessment either with or without accommodations (as specified in each student's IEP) in the state's general assessment or through an alternate assessment (Roeber & Warlick, 2001). IDEA further required results from the alternate assessments to be aggregated and reported with results of the general assessment for accountability purposes.

The federal assessment requirements contained in IASA and IDEA forced states to take a more active role in their state assessment programs. The requirements also made it much more difficult for states to rely solely on a commercial, off-the-shelf NRT as their state assessment. Any remaining questions about the appropriateness of relying on an off-the-shelf NRT as the sole or primary state assessment instrument were answered in the 2003 U.S. Department of Education guidance on state assessments allowable to meet the requirements of NCLB:

> A State may include either criterion-referenced assessments or augmented norm-referenced assessments in its assessment system. States wanting to use a norm-referenced assessment at a particular grade must augment that assessment with additional items as necessary to measure accurately the depth and breadth of the State's academic content standards and the assessment must express student results in terms of the State's student academic achievement standards. (U.S. Department of Education, 2003, pp. 13–14)

All of the factors discussed in this section contributed to changing the nature and role of state assessment and, consequently, redefining the relationship between states and testing companies by requiring states to assume a more active role in their assessment program. This is reflected in annual surveys of state assessment programs conducted by the Council of Chief State School Officers that show significant increases from 1996 to 2001 in both the number of full-time, professional state department staff dedicated to assessment activities and the annual budgets for state assessments (Council of Chief State School Officers, 1996, 2002). Increased assessment demands imposed by NCLB will continue this trend. Ownership of state assessment programs was transferred from the testing company to the state— literally in many states, at least figuratively in almost all states.

Summary

The need to track assessment results from year to year and the shift of ownership of the state assessment from the testing company to the state did more than simply add complexity to the process of transitioning between testing contractors. It is not an overstatement to assert that those factors created a need for a systematic transition process that did not exist previously. In an earlier era of state assessment, there was little, if any, need for a new testing contractor to have more than a cursory understanding of the procedures of the previous contractor. In the current era of high-stakes assessment and accountability, the state assessment program cannot function unless the new contractor can successfully implement the program's existing policies and procedures.

Understanding this change and recognizing that a transition process is a new and necessary component of a state assessment is a critical first step in achieving a successful transition. Unfortunately, it is often easy to overlook new steps in a complex process that is in the midst of transformation. Testing companies are focused on transforming organizations built to produce and market off-the-shelf, norm-referenced tests to organizations that can meet the demand to provide services for custom state assessments. State assessment directors are faced with learning on-the-job, in a highly charged, political environment, about how to be CEO, COO, and CFO of a multimillion dollar state assessment program. The need for a transition process may not become obvious to either party until it is too late—often when the state is scheduled to report the results of the first assessment administered by the new contractor.

Managing a Successful Transition in the Short-Term—Key Elements

Having acknowledged that a transition process is necessary, the state must be prepared to manage a successful transition. The first step is to recognize that the transition process is a separate project from the day-to-day operation of the assessment program. Like any other project, managing a transition begins with identifying the purpose or goals of the transition. One state assessment director summarized the goal of the transition process as ensuring that she would be able to stand before students, parents, the public, and the press next fall and declare that any changes in assessment results, positive or negative, were caused by changes in student performance and not by the change in testing contractor. Accomplishing that goal requires developing a comprehensive project plan, allocating appropriate resources (i.e., money, time, and staff), implementing the plan, and monitoring results.

Although the specifics of the transition process will vary across assessment programs, there are key components to a successful transition that all states should consider: *planning for the transition, speaking a common language, and clarifying expectations.* The common thread across these components is communication.

Planning for the Transition

Planning for a transition between testing contractors is a process that begins long before it is time for the state to issue a new RFP and face the prospect of hiring a new testing contractor. In fact, key parts of the transition process should have been in place before the old contractor was hired. Plans for how the relationship with a contractor will end should be communicated and agreed upon before the relationship begins to increase the likelihood of a successful transition to the next contractor.

At a basic level, planning for a transition can be similar to crafting a prenuptial agreement. Going into the contract, there must be clear understanding on what is owned by the state, what is owned by the contractor, and what each party will take from the contract. At a time when there is no discord between the two parties, they should reach agreement on the closeout activities that will occur at the conclusion of the contract—whether that conclusion occurs at the scheduled end date or at an earlier point in time. There should be no dispute on issues such as the need for the prompt and complete transfer of materials, the expectation that the contractor will cooperate and assist with the transition, and the level of interaction, if any, that is expected between the old and new contractors.

At a deeper level, however, planning for successful transitions requires regarding the state assessment program as an institution that will outlast the individuals who are involved in its operation at any particular time—much like the state government or even the testing company. Systems should be implemented to ensure that all of the information and materials needed to support the ongoing operation of the program are collected routinely, maintained, and are accessible. Checks and balances should be established to prevent any single individual or group of individuals from inflicting too much damage on the institution.

It may seem overly optimistic (or pessimistic, if you prefer) to discuss state assessment programs in the context of institutions built to endure. With some notable exceptions, many state assessments have had life spans similar to the firefly or butterfly. Nevertheless, with increased stakes and an emphasis on measuring growth and improvement over time there is a critical need to establish systems designed to protect the assessment program from an over-reliance on specific individuals from either the state or the testing contractor.

Speaking a Common Language

As difficult as it may be to believe after spending any time in conversations filled with acronyms and arcane terms, the testing industry lacks a jargon—in the sense of a shared, specialized, technical language that is common across the industry. It would be more accurate to say that the testing industry has dialects—regional (i.e., company or state) variations of a standard language in which vocabulary and speech patterns differ. Like regional dialects in the traditional sense, these language variations in the testing industry may be caused, in part, by physical separation or isolation. In the company-based NRT testing era described previously, there was little need for interaction among the relatively small number of large testing companies. It was easy for a company to function with its peculiar internal ways of speaking.

At the beginning of a contract, most states and contractors are prepared for the normal, new relationship phase that includes getting to know and trust staff and understanding the other's organizational culture and structure (i.e., work flow, approval streams, methods). A large part of this initial period is learning new names for familiar processes, operations, or materials. Working with a new contractor (or state) can require learning a new language and it may be necessary to develop a Company A to Company B dictionary—one company's *pop* is another company's *soda*. Perhaps even more important than learning new names for familiar processes, states and testing companies must ensure that they have the same expectations when they are using the same terms. As discussed in the next element, Clarifying Expectations, states and contractors may be using the same terms but have very different expectations for the product or service to be delivered.

Clarifying Expectations

Closely related to the process of developing common terminology is clarifying the state's expectations. The state and contractor must arrive at a clear understanding of what the state's specifications are for all products and services that the contractor will deliver. Consider a relatively simple, but critical, product such as an *anchor set* for scoring. How many papers does the anchor set contain at each score point? Do all of the anchor papers reflect the prototypical response or are there examples of unusual responses or borderline papers? How much involvement does the state expect to have in the selection of individual papers and approval of the set?

Although it is common practice to include as much detail as possible in the request for proposals (RFP) for end products (e.g., test booklets, manuals, reports), it is just as important to specify the details of internal or process deliverables such as anchor sets in the RFP, if at all possible, or at the latest, in the contract negotiation. The consequences of not clarifying these expectations can be enormous. On one level, there is always the ill feeling generated when a deliverable does not meet the state's expectations and the contractor's staff knows that "this is the way we usually do it." Of course, there are also budget implications when the contractor has budgeted to deliver an anchor set containing five sample responses (one per score point) and the state expects to select a final anchor set of 25 responses (five per score point) after having the opportunity to review 50–75 responses at each score point. Multiply the additional costs for an individual anchor set by the number of items in each test by the number of tests; add in additional costs for photocopying, processing, and shipping multiple sets of materials to the state; and a misunderstanding in expectations can lead to either a large expense absorbed by the contractor or a large change order presented to the state—neither of which is an appealing option.

Summary

The three elements described in this section—planning for the transition, speaking a common language, and clarifying expectations—represent the minimal, initial steps that must be taken to improve transitions between contractors. Some states have begun to incorporate these elements into their RFP process, contract negotiations, and the daily operations of their state assessment programs. We are still, however, at a relatively early point in the current era of state-owned state assessment that emerged in a few states in the early 1990s, began to take shape slowly in the late 1990s after the passage of IASA, and only reached its tipping point with the passage of NCLB in January 2002. Many states have not yet completed a transition in which it was necessary to maintain existing achievement standards across grades 3 through 8 and high school with new tests. Other states have endured their first unplanned transition in which they have had to face issues such as disagreements over ownership of existing items, problems with implementing consistent scoring and psychometric procedures, or the necessity to negotiate supplemental one-year contracts with their existing contractor and build special linking studies in an attempt to maintain achievement standards. In short, we are at a point where there is increased awareness of the need to plan for transitions between contractors but there is not yet a widespread, deep understanding of all that is required for a successful transition.

The following section provides a brief case study of how one state applied lessons learned from an unplanned transition in 2000 to prepare for and manage a much smoother transition between contractors in 2004.

A Case Study in Transitions—The Massachusetts Comprehensive Assessment System

Background

The process of learning through experience to manage the transition process is illustrated by the difference in approaches between the Massachusetts Department of Education's 2003 and 1999 issue of a request for proposals (RFP) for the Massachusetts Comprehensive Assessment System (MCAS). In 2003, the Department prepared to issue an RFP for a multiyear contract to assist in the administration of its state assessment program. The contract awarded under this RFP would be the third five-year contract awarded under the current state assessment program. The first contract period (1995–1999) covered the design of the assessment, pilot testing, and the first two operational administrations of the assessment. The second contract period (2000–2004) covered the next five operational administrations of the test. The third contract period (2004–2008) would cover the continued operation of the existing tests and the development of new tests to meet the requirements of NCLB.

The First Transition

In 1995, Measured Progress (known at that time as Advanced Systems) was selected as the state's contractor for the first MCAS contract. As a result of the second RFP process in the winter of 1999–2000, the state assessment contract was awarded to a new contractor, Harcourt Educational Measurement, and the state was faced with an unplanned transition. At that point, the state had never been involved in transitioning an existing testing program to a new contractor and had not anticipated the complexity of the process. Under existing contracts at that time, there was no overlap between the two assessment contracts. In fact, due to delays in the RFP process, the current contract had expired before the new contractor was selected. The spring 2000 administration of the state assessment was scheduled to begin approximately 75 days after the new contractor was selected—well before it would be possible to negotiate and finalize an official contract with the new contractor. Additionally, no plans or contractual agreements were in place to transfer information from Measured Progress to Harcourt.

Ultimately, a transition plan was crafted on the fly and the state assessment was administered under less than ideal circumstances that required the implementation of several one-time-only policies and procedures. Test booklets were produced and printed by Measured Progress. Student answer booklets, however, were produced and printed by Harcourt, then packed and shipped in bulk to Measured Progress (a shipping process very different from normal Harcourt operations). All test materials were then shipped to schools by Measured Progress. At the conclusion of testing, schools were directed to return test materials to Harcourt. Attempts were made to coordinate the shipping and receiving procedures of the two companies and to direct questions from schools to the appropriate contractor. Measured Progress's role in the 2000 MCAS administration officially ended at the conclusion of the test administration window. Approximately 100 days after being selected as the testing contractor, Harcourt was attempting to process, score, and equate tests while simultaneously sifting through piles of transferred hard copy documentation and files to determine the existing policies and procedures. Ultimately, after a great deal of unanticipated and unbudgeted work by the state and the contractors, as well as several negotiations of revised schedules, score reports for the 2000 MCAS administration were issued in late November 2000—approximately two months later than anticipated in the original schedule.

The impact of the unplanned transition between contractors did not end, however, with the reporting of results from the 2000 MCAS test administration. At the same time when processing of the 2000 MCAS tests was occurring, planning and development of the 2001 MCAS tests was underway. The transfer of test items and related materials provides a representative example of the issues that were faced throughout the program. With the exception of newly field-tested items, all items appearing on MCAS test forms in a given year have been previously administered in a prior year. Because of the hasty transition and the incompatibility of systems between the two contractors, virtually all of the previously administered items had to be recreated from scratch without being altered in any way—including attempts to match fonts. Additional work was also needed to adapt all of the identifying information and historical data associated with the items into Harcourt's data management systems. The labor and time needed to complete these tasks greatly impacted the schedule and process for reviewing items field-tested on the 2000 MCAS test and ultimately impacted the schedule for developing new items to be field-tested on the 2001 MCAS tests. In addition to those technical issues, there were also many personal and personnel issues to resolve as Harcourt began to fully under-stand the state's expectations and staff from the state and contrac-tor established working relationships under tight deadlines and stress-filled circumstances.

Learning from History

To avoid repeating the chaotic transition that occurred in 2000, the state's third RFP included a one-year overlap between the current and new assessment contracts. In part, the overlapping year was included to allow development of new tests being added to the state assessment program to meet the requirements of NCLB to begin. Significant resources, however, were dedicated to the possible need for a transition between contractors. As fate would have it, Measured Progress (with CTB/McGraw Hill as a subcontractor) was selected as the contractor for the 2004–2008 contract period and a second transition involving Measured Progress, Harcourt, and the Massachusetts Department of Education was required.

Ultimately, staff from the state and contractors totaling 3.0 person years and a budget equivalent to approximately 15% of the annual assessment budget was allocated to the transition. In contrast, staff time devoted to the prior transition totaled 0.25 person years and no funds were specifically designated for the transition. Key components of the transition plan were the development of a transition team and the scheduling of a series of transition meetings with the state and its two contractors. The transition team included senior assessment staff from the state as well as project management resources from each of the contractors. Over the course of the year, seven multiday transition meetings were scheduled, with each meeting dedicated to a major component of the assessment program: General Orientation and Goals for Transition; Master Project Schedule and Project Plan; Data Collection, Scanning, and Processing; and Publications; Analysis and Reporting; Psychometrics; and Scoring. The purpose of each of the meetings was to fully describe the procedures and processes related to that component, identify the materials to be transferred between contractors, and determine the specifications for the transfer of materials. Transfer specifications included the format of materials, schedule for delivery, method of transfer, identification of staff responsible for transfer, and criteria for verifying a complete and accurate transfer.

Another critical aspect of the transition was the planning and implementation of a series of replication studies. The purpose of these studies, designed with the assistance of the state's Technical Advisory Committee, was to demonstrate that the Measured Progress and CTB/McGraw Hill could replicate the results produced by the Harcourt in three technical areas: scoring of constructed-response items, item calibration/scaling, and the equating of tests across years. Similar studies were conducted following the previous change in contractors. However, because most of the analyses were not completed until after the formal transition period had concluded, there was no formal mechanism in place for the new contractor, or the state, to work with the old contractor to answer questions and resolve discrepancies that arose. The extended duration of the second transition removed this barrier and significantly simplified the process of conducting the replication studies.

Improving the Transition Process—A Vision for the Future

The focus of the previous section was on preparing for and managing a successful transition between contractors under current conditions characterized by limited standardization and communication across the testing industry. A great deal of time and effort in successful transitions is devoted to determining how key information, data, and products developed by one company can be efficiently formatted and transferred for use by another company. For future transitions to be considered successful, the need for much of this discussion must be eliminated.

Flattening State Assessments

In *The World is Flat*, Friedman (2005) describes how a key to the ongoing technological revolution was consumers demanding and technology companies responding to the demand for seamless interoperability between the computer systems and software applications developed by different companies. While participating with representatives from three state assessment programs and two testing companies in the symposium that provided the foundation for this chapter, a perfect example of this seamless interoperability played out. In the midst of the session, the laptop computer on which all of the presentations were stored stopped functioning. While the presenter continued speaking, presentations from the remaining presenters were collected from different USB drives and loaded onto a new laptop computer (a different brand than the original laptop). The new laptop was connected to the LCD projector and the symposium continued—all without restarting a computer or turning off the projector. No time or attention was needed to consider what companies developed the USB drives, laptop computers, presentation software, or LCD projector.

There is a similar demand for this level of interoperability in the testing industry today. The basic element of the test, the item, serves as a prime example of this need. A typical test item consists of for- matted text and a graphic. Test items may be developed years before their use on a test and then may be used operationally for several years. At a minimum, test items developed by a previous contractor must be easily accessible for inclusion on a test constructed by the new contractor. There should be no need for the new contractor to reproduce and reformat the items for use in their applications. In fact, requiring the new testing contractor to reproduce the test item only introduces opportunity for error into the system.

Populating the test instrument itself, however, is only one of many uses for the test item. In addition to the test, a state's assessment unit may have developed or purchased an item bank application for the storage of items or for the use of released items in formative, classroom assessments. The state needs the ability to import items from any testing contractor into its item bank application. Beyond the assessment unit, the state's information management unit may maintain a data warehouse and analysis system that includes direct links to specific released items. The data warehouse must be able to accept items prepared by any testing contractor. The test item, of course, is only one of many products and procedures that must be seamlessly transferred during a transition in state assessment contractors.

The current level of interoperability did not exist in the com- puter industry a few years ago and did not occur by chance. It is the result of technology companies, large and small, working together to develop common standards for the development of the fundamental tools and information that consumers wanted and needed to share without regard to platform or software application. The problem, therefore, is determining how to achieve this level of interoperability within and across state assessment programs.

The first step to developing common standards is for states and testing companies to reach agreement on the common elements of state assessment programs that must be transferred during any transition. At one level, this might include highly visible elements of the program such as test items (if they are owned by the state), scoring materials for constructed-response items, manuals, test booklets, and ancillary publications. At another level of program operation, the list of common elements to transfer among states and companies might include item parameters, item statistics, scoring specifications, equating specifications and procedures, and processing rules and procedures established for the assessment program. Simply identifying all of the fundamental components of state assessment programs will be a major task and a major accomplishment.

After the common elements have been identified, the next step in the process will be to develop common standards. Experience in other industries has shown that this exercise must be directed toward meeting consumer demands and must be voluntary and include all stakeholders. In the context of state assessment, the primary players are the state departments of education and the testing companies. The requirements, however, and capacity of end users of state assessments such as districts and schools must also be considered in developing common standards. Major support industries such as shipping companies and printers must also be included in the process. The process is based on openness, consensus, and compromise and cannot become dominated by larger testing companies focused on protecting their turf. Fortunately, the states and testing companies will not have to invent the procedures for developing common standards for state assessments. There are established procedures for developing common industry standards and organizations dedicated to the guiding industries in the development of voluntary common industry standards (American National Standards Institute, 2005).

The development of common standards to ensure that the fundamental systems of state assessment programs are interoperable does not mean either that all state assessments will look the same or that all testing companies will be required to standardize internal procedures and operate in the same way. The goal of common standards is to ensure that the pieces of the programs that must be transferred among states and companies are easily exchanged in a standard manner. The innovative and proprietary processes and procedures that individual companies develop to produce high quality test items, efficiently track and monitor materials, score student responses, or generate test results will not be threatened by common standards. In fact, many of the resources now devoted to interpreting nonstandard materials received from states and other testing companies can be redirected toward improving internal processes and procedures.

Conclusions and Lessons Learned

Changes in the characteristics, uses, and ownership of state assessment programs have fundamentally changed transitions between testing contractors. Organizational structures and operational systems designed to support the previous era of state assessment programs are not adequate to meet the current needs. There is a need for short-term, immediate solutions to help states and contractors successfully manage these new transitions. The increased scope of state assessments and the high stakes associated with the results of state assessment make it increasingly difficult for the state-contractor relationship to survive a poorly managed transition. Short-term solutions such as planning for transitions through overlapping contracts, increased communication between contractors during the transition, and more specification of contract requirements can increase the likelihood of a successful transition. Some states and testing companies have begun to recognize this need and are incorporating transition planning into the contract process.

In the long-term, however, effective transition planning will not be a sufficient solution. Systemic changes are required in the management and operation of state assessment programs by states and testing contractors to ensure seamless transitions between testing contractors and the efficient operation of state assessment programs. Achieving this level of interoperability in the state assessment industry is a process that has not yet begun.

Notes

The material presented in this chapter builds upon a symposium presented at the 2005 Council of Chief State School Officers National Conference on Large-Scale Assessment in San Antonio, Texas. The author thanks colleagues from state departments of education and testing contractors who organized and participated in that symposium: Cornelia Orr, Kit Viator, Mikel Brightman, Lisa Ehrlich, and Charlene Tucker.

References

American National Standards Institute. (2005). *Overview of the U.S. standardization system: Understanding the U.S. voluntary consensus standardization and conformity assessment infrastructure.* Washington, DC: Author.

Council of Chief State School Officers. (1996). *Annual survey of state student assessment programs.* Washington, DC: Author.

Council of Chief State School Officers. (2002). *State student assessment programs: Annual survey. (1).* Washington, DC: Author.

Miller, L. (1997). *Quality Counts 1997: A report card on the condition of education in the 50 states.* Washington, DC: Editorial Projects in Education.

Friedman, T. (2005). *The world is flat: A brief history of the twenty-first century.* New York: Farrar, Straus & Giroux.

Hurst, D., Tan, A., Meek, A., & Sellars, J. (2004, April). Overview and inventory of state education reforms: 1990 to 2000. *Education Statistics Quarterly.*

Improving America's Schools Act of 1994. (1994). *Title I—Helping disadvantaged children meet high standards. Section 111.* Retrieved October 18, 2005, from http://www.ed.gov/legislation/ESEA/toc.html

No Child Left Behind Act of 2001. (2001). *Title I—Improving the academic achievement of the disadvantaged. Section 1111.* Retrieved from http://www.ed.gov/policy/elsec/leg/esea02/107-110.pdf

Olson, L. (1999). Making every test count. *Quality Counts '99.* Washington, DC: Editorial Projects in Education.

Roeber, E., & Warlick, K. (2001, Autumn). Challenge and change of IDEA '97. *The State Education Standard,* 8–13.

Rothman, R. (1995). *Measuring up: Standards, assessments, and school reform.* San Francisco: Jossey-Bass Publishers.

U.S. Department of Education. (2003). *Standards and assessments: Non-regulatory draft guidance.* Washington, DC: Author.

11

Department of Education Perspective on Effective Vendor Relations

Judson Turner
Executive Counsel, Office of the
Governor, State of Georgia

Introduction

With respect to statewide assessments at the elementary and second-ary levels, the primary objective for both the vendor and the govern-ment should be the successful development, administration, scoring, and reporting of valid and reliable tests. In order for local educational professionals and parents to utilize testing data and thus make the best educational determinations for the student, assessments must not only be developed appropriately for utilization by the end user, but these assessments must be administered, scored, and reported according to a timeline that leaves little room for error. As general counsel to the Georgia Department of Education, I learned a great deal about large-scale assessments and the legal and organizational challenges that must be addressed in order for those assessments to be successfully utilized at the state and local levels.

This chapter explores those tools necessary in the development of a successful statewide assessment program built in part upon the outsourcing of large assessment contracts to private sector vendors. As with any profitable business enterprise, the right people must be employed in the right jobs with the right institutional support in order to achieve success. Thus, the first and foundational topic to be covered surrounds the creation and support of a quality testing divi-sion within the state department of education. Second, we cover the essential development of a tight, well-written request for proposal, including the typical pitfalls that befall many public procurements and useful techniques for avoiding those pitfalls. Next, we look at the culmination of the public procurement—the successful negotiation of a strong contract, providing the requisite terms and conditions of the relationship. Finally, we address the management of that contract and the organizational features required for that task.

Step 1—The Creation and Support of a Quality Testing Division

Building a successful statewide assessment program begins with the state partner. More specifically, the single most important action that can be undertaken in order to build an effective assessment program is the organization, development, and support of the testing division within the applicable state department. Whether one approaches this topic from the vendor or state perspective, the task of delivering a well-developed and well-administered assessment hinges in significant part upon the quality of the state partner. Although there may be apparent advantages from the vendor perspective to a hands-off or delegated approach to assessment contract development and management, more often than not, such an approach leads to the development of unforeseen and often more pernicious problems.

I assumed the role of general counsel to the Georgia Department of Education in early 2003, having started my law practice as a commercial litigator in a large law firm. When I arrived at the Georgia Department of Education, the legal division and a testing division did not view themselves as necessary, much less vital, partners. Even more disturbing was the separation between the testing division and the rest of the Georgia Department of Education. The result of this dysfunctional organization manifested itself quickly. Within two months of my arrival, the award of the multiyear public procurement for the state's foundational assessment for grades 1 though 8 (the Georgia Criterion Referenced Competency Test, or CRCTs) had been overturned in response to a bid protest. The protest was granted because of a minor, but indisputably important, technicality.

By way of historical background, Georgia had administered the CRCT for English/language arts and math in grades 4, 6, and 8 since spring of 2000. Beginning with the 2001–2002 school year, Georgia expanded the CRCT to science and social studies in grades 3 through 8 and grades 1 through 8 in English/language arts and math. In the fall of 2002, the CRCT statewide assessment contract for English/

language arts, math, science, and social studies for grades 1 through 8 was awarded to another vendor under a six-year, annually renewable contract. Due to factors that we discuss in greater detail in this chapter, the Department's RFP was entirely too broad, resulting in bids from the $60 million range to the over $120 million dollar range. After the technical scores were tallied, only one vendor had received the technical score sufficient to advance to have its cost proposal reviewed and contract negotiated. Although the high-cost vendor was the only vendor that made the requisite technical score, upon review of the actual cost, the proposal from this vendor was prohibitively high for the state.

This development left the state in the unenviable position of trying to negotiate down the cost during contract negotiations without reducing the scope of the RFP itself. Unfortunately, while the state was successful in shaving over $20 million off of the high-cost vendor's proposal, it did so at great risk under the procurement rules. The moment the contract was awarded, a bid protest was filed alleging, among other things, that the contract negotiations had impermissibly reduced the scope of the contract. With roughly $20 million of cost reduced, the department was not able to prevail in its position that the original scope of the RFP had not been impermissibly reduced. It was upon this ground that the contract was overturned, leaving the state to negotiate an emergency contract with less than three months to go until the spring administration of the CRCT.

Even more problematic than the contractual problems associated with the contract award, the bank of test questions from which the state's operational assessments and its online practice questions were drawn had been compromised. As is often the case in the world of public procurement, the state experienced the challenge of dealing with multiple vendors in order to provide the CRCTs in successive years. This reality meant that different vendors manipulated the state's test question bank at different times. Through complications resulting from this and a combination of other factors, questions from the test bank's secure pool (to be used for building the operational test forms) were compromised.

Several of the secure questions from the test bank had been viewed during the school year by teachers and students using the state's online practice question functionality. Although the causes for this exposure were not fully understood, part of the problem resulted from the coding of these questions within the test question bank. Secure questions were to be pulled for use on the operational tests by a computer program designed to read the coding of the test questions. A separate set of questions, coded at a different level, were to be made available for online usage by teachers and students in preparation for the operational assessment. Somehow, teachers and students in preparation for the spring assessment had practiced, unbeknownst to them, on questions that were slated for inclusion in that spring's operational assessment.

Because of the inability to handle the extent of the exposures of questions included on the operational assessment through the post-equating process, the assessments in all but grades 4, 6, and 8 had to be canceled for that year. The state, thus, lost much of the valuable data that would have been gathered from the second year of assessments in grades 1 through 8. Moreover, the state spent significant taxpayer funds *cleaning* the test bank to recreate a secure pool of questions for the assemblage of future operational forms. Such costs, however, pale in comparison to the additional cost Georgia realized when it returned to the market to procure a multiyear contract for the CRCTs under such circumstances. Although testing contracts seem to be on an ever-increasing inflationary trajectory, no one will dispute that Georgia had to pay a premium to procure its CRCT contract given the risks and unresolved test bank problems at the time of the procurement.

One or more private sector testing vendors may bear legal responsibility for the problems resulting from this incident; nonetheless, in the end the Department of Education cannot avoid some of the blame.[1] No matter how well drafted the contract and no matter how good the reputation of the state's testing contractor, there is no substitution for a knowledgeable, engaged, and well integrated state partner. Without such a state partner, the enterprise of building that quality statewide assessment program is one wrought with additional challenge and difficulty. Thus, the first step toward realizing that goal is to build a well-integrated and knowledgeable testing division in the state.

Technical Expertise

Having a testing division that contains sufficient technical expertise is certainly a foundational characteristic of a successful testing team. As I learned early on in my experience negotiating and managing large assessment contracts, the development and administration of large-scale assessments are complex tasks, and a testing team, anchored by qualified psychometricians that can interface on both the technical and administrative sides, is invaluable. Given my admitted limitations on the technical side, I do not attempt to provide any further instruction on the necessary technical expertise. Suffice it to say, anyone with control and responsibility for a testing division should be willing to learn from those successful testing divisions and the requisite education and experience levels of those staff.

My comments regarding the requisite technical expertise needed to build a quality testing division should not be taken to suggest that technical expertise alone guarantees success. Rather, as we shall discuss, any quality testing division should contain or have ready access to the services of those skilled in areas outside the technicalities of test development and administration.

Organizational Integration and Support

Integration within and support from the greater governmental department are vital to the development of a successful testing division and assessment program. Many of the tools and techniques we discuss later in this chapter require the utilization of different skill sets, particularly in the areas of contract development, negotiation, and management. A good working relationship between the applicable legal and policy divisions and the testing division is imperative. Failure to include within the working team that legal and policy expertise will invariably lead to decisions which fail to take into account all the political, policy, and legal ramifications inherent in these programs, thus creating problems beyond the natural challenges that come with large-scale assessments.

Even for those that may approach this issue from the vendor perspective, it is important not to skip over the first step of the organizational composition and institutional support of the state department's testing division. Although the vendor may not control the development and quality of the state's testing division, understanding where those deficiencies lie will be integral to building a vendor team that may be able to compensate, at least in part, for some of those deficiencies and, thus, allow the venture to achieve a successful end result. This observation is particularly true where the testing division appears to be isolated from the policy development and initiation divisions within the department. Knowing when to confirm operational decisions that have wide ranging policy implications can be invaluable. Where a testing division seems to lack the optimal organizational integration and respect within the greater governmental department, a vendor having a direct alternate access to a higher official within the state organization may prove essential.

Step 2—Designing the Tight RFP

Before developing a quality, customized assessment contract to govern the relationship, one must first build a quality procurement solicitation document or request for proposal (RFP), as they are commonly called. This task is far more difficult than it at first appears.

The Challenges Confronting the Development of Quality Procurement Solicitations

Because of the public nature of most statewide testing contracts, several complicating factors are present.

First, most states struggle in their development of RFPs of all kinds, including assessment contracts, because they don't often know exactly how much money they have to spend. At best, the state may only know its assessment budget now and the preceding fiscal years. Although this information may be of some predictive value for future years, not knowing exactly what kind of resources are available will undeniably complicate the procurement process as the policy makers struggle with the different choices.

Second, building a successful RFP requires significant preplanning and thought. Often in government bureaucracies, roles can be ill-defined and, as is especially true during leaner budget times, divisions may remain understaffed. Both of these factors result in lack of preplanning and thought that in turn lead to rushed RFPs. These rushed RFP's are often not customized for the needs of that particular state and fail to provide adequately for the resolution of future changes and developments.

Third, human nature propels one to do things as they have been done before, especially if the previous results were positive. This phenomenon is especially prevalent within government. Such an approach, however seemingly innocent, can be devastating in the public procurement arena. Foremost among the mal-effects of this tendency is the inclination to regurgitate previous procurement documents without giving adequate new and fresh consideration to the procurement at hand. Nothing could be more devastating to the development of a quality testing contract than such an approach.

Fourth, each state has slightly different procurement rules and a different legal environment. That which may be legal in one state may not be legal in another or may not be interpreted in the same way by the courts of another. Thus, whereas the procurement counsel and anecdotal advice from one state may be somewhat helpful, there is no substitution for becoming intimately familiar with the law and legal precedents relative to those procurement laws in the state in which you are engaged.

Fifth, given the transactional costs associated with public procurements, the rational approach suggests the solicitation of long-term, multiple-year contracts that do not have to be rebid often. Multiyear procurements may, in fact, save states significant resources in the long term and allow states not only to avoid the transactional costs associated with public procurements, but realize the economical efficiencies that come with stable, long-term vendor relationships. One pitfall is associated, however, with this approach. Often a state will be tempted to throw into its solicitation a couple of long-term functionalities that the state is not presently able to accommodate, but that the state may be interested in purchasing in the future. I am not suggesting that such strategic planning is in all cases inappropriate or unwise. Rather, I am suggesting that the procurement of such future functionalities may require significant consideration and planning and, moreover, may require a significantly more sophisticated RFP and resulting contract.

For example, instead of asking for certain functionalities to be included in the cost proposal whether or not the state will realistically ask for them during the life of the contract, the RFP might call for the ability to provide that functionality in the out years, with some arrangement for how that additional functionality will be paid for should the state elect to receive that deliverable. Instead of putting the vendor in the difficult position of deciding whether to build into its present cost proposal that which may never be utilized, this method allows the vendor to bid based on those mandatory deliverables contained in the RFP, while setting up a process with cost limitations for the negotiation of future identified deliverables without the requirement that the procurement be rebid at that later date.

"What" to Procure

Although each of these challenges undeniably influence the government agencies tasked with the job of developing quality procurement solicitation documents, I found one insight regarding them to be incredibly helpful and worth repeating in this context. Public procurements are far more art than science. Although there are certain rules that must be followed from a legal process standpoint, how one designs the procurement and the relative priority placed on the different deliverables of the RFP can only properly be accomplished through creative, multidimensional consideration and planning. One must consider each of the deliverables associated with a large-scale assessment contract and the relative importance one places on each of these deliverables. For instance, prompt reporting of test scores, even in this post–No Child Left Behind world, may vary in importance between states. Another state may be quite interested in developing new technology surrounding testing, whereas yet another may be driven more principally by cost considerations. Each of these factors are just examples of considerations that are undeniably present for all state testing divisions. How they relate together and the relative importance they receive in the RFP and resulting contract is a multidimensional consideration. Such consideration will take significant time and energy during the RFP development stages by the state testing division in concert with its applicable policy and legal partners.

Before any procurement team can help develop the right RFP, that is, *how* to procure, the policy team in concert with the technical staff of the testing division must settle precisely on *what* to procure. Determining what to procure, however, sounds far simpler than it is in the real world of statewide assessment contracts. Given the influencing factors mentioned previously, states will often want a series of options regarding additional functionality in out year contracts. Of course, everything comes with a price, and options regarding additional functionalities are no different. Online testing is a superb example of this type of additional functionality. Thus, the development team must expend resources determining this condition of the state relative to the need and this ability to utilize such an additional functionality. Next, the team must conduct an analysis regarding the future ability and desire for the state to utilize the additional functionality. Such groundwork and analysis must first be conducted in order to best determine whether that additional functionality ought to be included in this procurement. Only then, after determining exactly *what* to procure, may the team develop the methods to achieve that result.

The Cost of Ill-Defined and Poorly Planned Procurement Solicitations

There can be no question regarding the general inclination toward and potential benefit of longer, multiyear assessment contracts. The continuity and economies of scale benefits are often significant. If, however, such an inclination leads to the development of a *bloated* RFP, any benefits of a multiyear contract will quickly be outpaced by the costs of such an arrangement. As mentioned previously, each procurement needs to be designed individually. For this reason, whether or not multiyear contracts are always the most prudent course is a difficult question. Certainly, the continuity and economies of scale benefits of multiyear contracts lead one to start with the assumption that such an arrangement is preferable; however, given the individual circumstances facing that particular state partner, a single-year contract may actually prove most prudent.

Not only does the failure to build a tight RFP cost the state in terms of the extra expense for functionality that the state may or may not use, but such an RFP will not accurately reflect the state's needs and relative priorities. Given the relative inflexibility of the public procurement process, a state may end up partnering with a vendor that may have won the public procurement based on technical *points*, but which does not share the state's relative priorities regarding the deliverables in the contract. In other words, these low-priority functionalities, if included in the RFP with too great a weight, may allow a vendor that would not otherwise be competitive to win. This unequally yoked partnership will often develop when the state fails to carefully design its technical scoring rubric in a manner that accurately reflects the priorities for that procurement. One might argue that such matters should be covered in the contractual arrangement between the parties. Contracts, however, often incorporate the RFP by reference, making the terms of the RFP part of the contract and thus perpetuating the ambiguities regarding such matters in the contract.

Furthermore, even if the parties are able to clarify such matters during the contract negotiation phase of the procurement, such protection can only go so far. The best procurements identify the best value for the taxpayer dollar. The best value is not necessarily the best price. Rather, best value really seeks to identify the proposal that offers the state the best quality product at the lowest price considering, not only the purchase price, but all the negative costs that will accrue to the state if the venture fails in any material aspect.

In reality, all too often RFPs are drafted more as instruments to test the waters of the market rather than tools to identify the right partner for the state's joint venture. RFPs that look more like Christmas trees with ornaments of every shape and color with varying technical points awarded for each are much more likely to elicit varying technical and cost responses. Vendors tasked with preparing proposals for such RFPs are left wondering what elements of the RFP are really essential to the state and which are just extras. Of course, if the RFP is particularly deficient, leaving vendors to believe that each ornament on the tree is essential, the cost of that contract will be proportionately more expensive. Either the state will find itself trying to sort through proposals that vary widely both technically and with respect to cost or the state will be left wondering where to find a cheaper Christmas tree.

Lastly, an ill-drafted RFP that produces varying technical and cost proposals will invariably lead to a confused partnership—one in which neither partner shares the same understanding of the relationship. Confused partnerships are recipes for failure and should be avoided at all costs. As noted, the best means of producing a clear partnership is through a tight, well-defined RFP and resulting contract.

Step 3—Contract Development and Negotiation

Having stressed the importance of a tightly defined RFP, some discussion regarding the actual contract between the state and its outside testing vendor is in order. Typically, contracts incorporate by reference the terms of the RFP with provision for any conflict between the terms of the contract and the RFP to be resolved in favor of the contract. After the RFP process is utilized to identify the correct vendor or the vendor whose solution and approach best meets the state's stated needs, it is the contract in conjunction with the RFP that governs the relationship between the state and its vendor going forward. The contract is the instrument that records the mutual understanding of the parties regarding the relationship and, depending upon its sophistication, provides steps for the resolution of conflicts that may develop during the life of the relationship. A good contract will use both carrots and sticks in order to best achieve its purpose.

Dealing with Contract Exceptions during Negotiation and Award

Before addressing specific provisions for inclusion in any statewide assessment contract, a couple of observations are probably appropriate. First, the initial draft contract is often included within the solicitation document. In Georgia, bidders must make specific exceptions to any of the provisions in the contract they wish to negotiate if they became the apparent successful bidder prior to contract negotiation. The issue of contract exceptions can be quite tricky during the procurement process. For instance, the procurement laws and regulations of many states do not provide for the evaluation of contract exceptions. This reality means that one vendor who has substituted its contract wholly for the state's contract, effectively taking exception to each of the state's contractual provisions, may receive the same technical score as the vendor that agrees to sign the state's contract with all its relevant provisions as drafted. Depending on the particular contractual provision at issue, however, a contract exception could have a significant effect upon the overall value of the proposal. One can quickly see that a $1 million cost proposal that takes exception to the performance bond provision of the contract is not the same value as a $1 million offer that makes no exception to this provision.[2] Of course, the actual cost of the performance bond to the vendor will depend on the financial stability of the vendor in addition to the performance record of the vendor and the relative risk of the project.

Although having the ability to evaluate contract exceptions might be viewed by some as the most comprehensive means of ascertaining the actual value of each vendor's proposal, leaving all the contract exceptions for the negotiation stage of the procurement also has its advantages. For instance, one may find that the actual contractual provision at issue is not worth the cost and, thus, having the ability to negotiate those terms after the apparent winner has been identified is a real plus. Moreover, nothing stated here should be taken to suggest that one must settle for a vendor that fails to agree to the terms and conditions the state requires in its contracts. Failure to agree to contractual terms provides sufficient grounds to move past that vendor to the next apparent successful vendor. Of course, before moving to the next vendor for failure to agree to terms and conditions in the contract, one should examine the contractual exceptions included in the proposals of the other vendors.

Although I have by no means solved the issue of how best to treat contract exceptions during the RFP award and contract negotiation phase, my goal was to identify the vast importance of contract exceptions to the public procurement process and encourage you to evaluate each provision before including it in the draft contract, understanding which of the contractual protections are negotiable and which are not.[3]

Second, much like the exercise that has to be initiated on the front end during the development of the RFP, one must realize that each of the potential provisions of any contract come with a cost.[4] I encourage the state to do as much work as possible in defining those nonnegotiable contract terms and, whether or not some contract exceptions are allowed, to make those essential terms clearly known as such in the solicitation document. Such an approach will hopefully avoid the uncomfortable scenario of being forced to move beyond the low qualifying bidder because the state refuses to compromise on certain contract terms that the apparent low qualifying bidder counted on negotiating away at the time it submitted its technical and cost proposals.

Contractual Provisions

I am sure you have often heard the old adage that *there is no such thing as a free lunch.* The adage certainly holds true with respect to contractual provisions. Many of the protections that we have successfully employed in Georgia came with a cost.

Payment Schedules   Many of the principal provisions that one ought to consider for utilization in statewide assessment contracts are by no means secret tools or groundbreaking tactics. In fact, much of what is involved in building a sound and strong contractual relationship is common sense. Designing a payment schedule that works for both parties is certainly a common sense tool.

Whether a product of convenience or stupidity, many statewide assessment contracts are paid in equal monthly installments—not based upon specific deliverables or with any percentage of the contract price held back until final performance. In the case of statewide assessment contracts, most of the significant deliverables are due in the last month or two of the fiscal year (which typically means in the last few months of that annual contract). Thus, paying the vendor in equal, monthly installments with no holdback leaves little leverage for the state at the end of the contract period, after as much as 5/6 of the contract price is paid out to the vendor by the time the main spring assessments are administered. Moreover, because the federal No Child Left Behind Act requires that Annual Yearly Progress determinations be made before the start of school for the coming year, virtually all states are trying to get results at the same time, leaving the states with smaller assessments or states with relatively weak contracts in a diminished position of leverage during this time.[5]

Having experienced the extreme leverage problem created by a contract with a payment schedule that failed to hold back any portion of the payment until successful final delivery and performance in Georgia, we were determined to build a payment schedule that fulfilled the following dual purposes: sufficient focus and stress upon important interim deliverables and successful final completion, including administration, scoring, and reporting of the spring operational assessments. We accomplished these objectives by varying the percentage payout based on specific deliverables and holding back 30% of the full contract price until successful completion of the administration, scoring and reporting of the spring assessments.

The creation of such a payment schedule provided the tools by which we managed the relationship during the performance of the contract; however, as noted previously, these contractual provisions did not come without some cost. How much extra we had to pay embedded in the overall bid proposal for the protection of a 30% holdback provision is unknown to me. Because of the cost of such a holdback provision, consideration should be given to the optimal

amount of a holdback. Georgia had had such a terrible track record
in previous years regarding the timely receipt of test scores that it was
willing to pay for a higher holdback percentage than what is more
commonly found in the industry.[6] Nearly all vendors that bid on
Georgia's CRCT contract tried to negotiate the holdback provision.
Our ability to successfully negotiate a contract with one of those vendors leaves me to believe that the cost proposals contemplated that
arrangement in some form, even if they did try and negotiate that
percentage down during contract negotiations. Moreover, I would
predict the state's firm position on a payment schedule fashioned
around deliverables and a stout holdback provision contributed significantly to the turnaround Georgia realized in the performance of
its CRCT contract.

Liquidated Damages   Liquidated damages are a popular tool for
use in contractual arrangements in which a timely delivery schedule is essential and damages for delay are difficult or impossible to
ascertain upon breach. We utilized liquidated damage provisions in
all our statewide assessment contracts. It should be noted, however,
that depending on the provisions of one's own state law, liquidated
damage provisions can be difficult to enforce.

Under Georgia law, both parties had to expressly agree on the
front end that in the event of a breach of the contract due to delay,
damages would be difficult if not impossible to ascertain. Therefore,
in light of that reality, the parties were agreeing on the front end to
a daily penalty payable by the vendor to the state until performance
was completed or cured.

Moreover, where liquidated damages are applicable, they tend to
be the exclusive contractual remedy. Thus, if one's actual damages far
exceed the amount received by operation of the liquidated damage
provision, courts will not typically allow one party out of that agreement to sue for actual damages. Again, the parties agreed on the front
end to liquidated damages because of the difficulty surrounding the
calculation of actual damages in this context; they should not be able
to maintain that damages are now miraculously calculable.

Notwithstanding all the disclaimers surrounding the use of liquidated damage provisions, their use can be extremely attractive in the
context of statewide assessment contracts. For instance, what would
one say is the damage to the state if the CRCTs are delivered one week
late to all the schools in the state and are thus scored and reported

a week later than planned? As noted earlier, such a situation might provide just the right context for the use of liquidated damage provisions. And, of course, it is not in their use that these provisions are most helpful. Rather, one hopes that they provide the right concrete consequence and incentive for the vendor to meet its obligations.

Performance Bond   Most state contracts of the size and importance of statewide assessment contracts include a requirement that a performance bond be purchased. Of course, the optimal amount of that bond varies depending on the size, risk, and relative importance of the project. As noted previously in my discussion of contract exceptions, vendors that seek to avoid or modify the required performance bond create problems.

First, depending on the size of the bond and the amount of the proposed reduction in the size of the bond sought by the vendor, the state may face significant limitations in the public procurement context. Any cost proposal included with a vendor response to the state's draft contract that includes an exception to the performance bond requirement must be carefully considered. Although other contract provisions negotiated away during the negotiation period may not have as identifiable a cost component, the cost of any given performance bond is certainly ascertainable and would thus arguably effect any cost proposal submitted by a vendor. One should carefully consider the procurement risks presented by agreeing to reduce a required performance bond during contract negotiations.

For instance, a competing vendor, particularly one that had a little higher cost proposal and did not take exception to the amount of the performance bond in the RFP, might convincingly argue in a bid protest that the contract award was improper given the agreement to reduce the amount of the required bond. Often those contractual terms and conditions most often negotiated do not ascertainably impact the vendor's underlying cost like the amount of a performance bond would. To avoid this whole procurement dilemma, I recommend giving sufficient thought and study to the right size for the performance bond and then making it a mandatory requirement of the RFP.

Second, performance bonds, depending upon their size, may be utilized in conjunction with or in substitution for some of the other contractual protections we have previously discussed, such as liquidated damage provisions. Obviously, a modest performance bond may fulfill one of the primary purposes for which they are utilized—to give the state some satisfaction regarding a vendor's financial stability and ability to perform. With sufficient planning, a modest performance bond may be best utilized in tandem with the other contractual provisions we have discussed to best ensure vendor success.

Alternatively, a large performance bond may serve as a sufficient substitute for some of the other contractual protections mentioned. Again, nothing comes without costs, and in return for a large performance bond the state might consider reducing the holdback provision or eliminating the use of liquidated damage provisions. Any attempt to reduce a performance bond or any failure to acquire the relevant performance bond will signal serious problems with the financial health of the proposed vendor. Thus, in either case, the performance bond requirement provides a real protection against those more risky private sector vendors.

Last, the actual terms of the required performance bond itself are worth considering and will also affect the vendor's ability to acquire the bond. Because the surety also has some say in whether the proposed vendor gets bonded, a careful analysis regarding the right size of the performance bond and terms of the bond itself is warranted.

Contract Management

Assuming the state has successfully completed all of the prerequisites we have discussed—creation of a quality testing division, development of a tight, well-written RFP, negotiation of a strong contract—contract management remains the last step toward the realization of a successful statewide assessment.

First, one must realize that contract management is a two-sided enterprise. Not only must the state be involved with the management of its private sector partner in order to ensure that the private sector vendor is responding appropriately to the inevitable challenges that develop during the life of a statewide assessment contract, but the vendor deserves the timely response from the state partner relative to those unforeseen developments.

Second, there is no substitute for timely detection and disclosure of performance problems or changes in realities on the ground. Again, this is a two-way street.

Third, as previously noted, many of the contractual protections we discussed before are not necessarily most helpful when the need arises to utilize them. In other words, although they place a consequence on poor performance in order to provide appropriate performance incentives to the vendor, in a world of growing demand and limited supply, when vendors nonetheless fail to perform, the contractual protections we have discussed often fall short in making the state whole in the case of a contractual breach. Although these provisions may provide money damages to the state, in the context of statewide assessment contracts, they will not get the test administered, scored, and reported.

Rather than just rely on the good will of the vendor and the performance incentives contained in the contractual provisions we discussed, there is no substitution for early detection and, as important, disclosure of performance problems, whether those problems be discovered by the vendor or the state. When those nonnegotiable changes in the realities on the ground occur, it is vital that the state sit down immediately with its vendor partners to discuss and agree upon the necessary course of action in response. Nothing in the contract should discourage either party from being anything but forthcoming regarding the status of performance and a partner in the required solution to any challenges that may arise.

One means to ensure prompt notification of developing challenges is to create a clear and open line of communication between the state and the vendor. This channel of communication can be established through weekly status conferences and periodic written updates. In my experience, both of these tools have been utilized successfully during the management of large assessment contracts. It should be said, however, that the vendor ought to be careful how it proposes to meet this obligation. In this technology age, long-distance status conferences via video conference or teleconference are often cost effective tools for this purpose. Nothing can frustrate a cost conscious state partner more than high administrative costs associated with travel by an out-of-state vendor. Additionally, both partners ought to balance the need to be updated with the costs in terms of time and money associated with such meetings. Some variance of the status conferences during different seasons of the contract makes good business sense for both parties.

I will take this opportunity to reinforce one of my earlier points regarding sophistication in the RFP and resulting contract. Proposals regarding travel costs and other administrative costs should be part of the procurement solicitation and evaluation processes. Expectations clearly laid out in the RFP and responses that clearly cover the treatment of those costs will lead to a much smoother contractual partnership—one that can be focused on the more threatening problems that will inevitably arise during the life of a contract of this complexity.

Once one has completed the hard work of building a solid testing team that can develop and design a sophisticated procurement solicitation document that lays out the state's needs and relative priorities and negotiate a strong contract, management of that contract is really the easy part. Without those prerequisites, however, that task becomes nearly impossible.

Georgia's Experience

As I mentioned at the beginning of this chapter, Georgia really faced some serious challenges within its testing program in the spring and summer of 2003. By and large, Georgia's testing program is past many of the foundational problems experienced in 2003. We were able successfully to procure the CRCT for grades 1 through 8, albeit at a distinct premium due to the problems with the test bank. We negotiated contracts that contained the provisions we discussed, and I believe those provisions, along with the general good will by Georgia's vendors to do quality work, contributed to the successful administration, scoring, and reporting of the CRCT in the years 2004, 2005, and 2006.

In 2003, we decided on a three-year renewal because we were not certain of what out year functionalities we might want associated with the CRCT, that is, online testing. Attempting to provide options for further out year functionalities was simply too costly in light of all the other risks associated with our test bank and testing program. Although the state will experience some of the transactional costs associated with a resolicitation in year 2006, I believe that a shorter procurement was the right course under the circumstances. I am confident that the risks associated with our program have decreased significantly, and, hopefully, that decrease in risk will translate into more competitive bids by more vendors this time around.

The state did expend significant resources to clean its test bank and is still trying to replenish many of the secure test questions that will be needed in coming years on its operational tests. That effort appears to have been successful, having conducted three years of testing without the exposures of 2003.

Due to the coding in the bank and the complexity required of the programming to fix these problems, remedying the exposure problem was no small task. We ended up hiring a technology vendor to build a program we needed to fit our internal system. This vendor did not have particular assessment experience prior to working with Georgia; however, the problems with the coding of the test bank were centrally a technology/coding problem as opposed to an assessment problem. A close partnership with the subject matter experts in Georgia's testing division was required in order for the state to be successful in this task.

Additionally, Georgia's Superintendent of Schools, Kathy Cox, in conjunction with the State Board of Education developed and initiated a wholly new curriculum in all grades and all four primary subject areas beginning in 2004. Thus, not only did the state work to rebuild its testing program, it did so while rolling out a new curriculum. The new curriculum, of course, required the development of new and more rigorous test questions. As the curriculum has been rolled out to different grades, Georgia's testing division has had to work in concert with its outside vendors to develop these newly aligned tests. So far, the new assessments appear to be working as expected—as students are tested on new content in a more rigorous fashion, thereby better preparing them for the competitive world that awaits them.

Finally, I should say that all our decisions regarding test development and procurement did not go as expected. Prior to 2003, the online hosting of practice tests and questions was under the same contract as the development, administration, scoring, and reporting of the operational tests. Given the problems encountered with the test question exposures and the theory that much more competition would be found for the online practice test hosting, we decided to break the traditional testing RFP and the online practice test hosting RFP into two separate RFPs. In the end, the same vendor won both procurements and the technology vendors that bid on the online hosting RFP were not as competitive as had been hoped. The state paid a significant premium for both contracts, but it is unclear whether the cost of both of these together was greater than what would have been the cost had we not kept them together.

As for actual contractual protections, I believe that the holdback provision played a significant part in the timely delivery of test data to Georgia and its school systems—where that had not occurred heretofore. Additionally, the performance bond requirement also resulted in the state moving past an apparent low bidder for failure to acquire the required bond. In the end, the bond allowed the state the flexibility not to contract with a company that lacked the financial stability and performance record to get bonded in the market, thereby reducing the risk of the procurement to the state. I am not aware of Georgia having invoked the liquidated damages provisions in any of its assessment contracts. As mentioned before, however, I believe the strength of these provisions is not necessarily in their utilization.

Outside of the technical areas of public procurement, Georgia's testing division has experienced a great deal of turnover and reorganization. Much of this type of reorganization is painful, as it was in Georgia. Generally speaking, Georgia moved toward the employment of a staff of both technical and management skill sets. In order to accomplish this goal, many existing employees either left the department or were relocated. As mentioned earlier, the charge to this group was to fix not only the problems of the past, but to design a testing program that fit a new era of ever rising expectations based in a new, more challenging curriculum.

All of this reorganization occurred during lean budget times when the state's assessment budget was deeply in the red. Certainly, there is a critical mass necessary within any state testing division in order to adequately manage and implement all the various state assessments; however, as is often the case in bureaucracies, it is hard to tell whether more employees are required or simply better employees. In Georgia's experience, we found that organizational integration, where the testing division had open access to the legal staff and the upper management of the department, allowed for the successful management and transition of large-scale assessments by a leaner staff.

Although not there yet, at the time of my departure, I believe that the testing division was far more integrated into the greater Georgia Department of Education and had direct access to the senior legal and administrative leadership of the Department. All in all, Georgia's testing story over the last few years is an encouraging one, and I hope one that continues to improve as Georgia looks toward the next round of large assessment procurements.

Conclusions and Lessons Learned

In summary, having inherited a rather tangled and potentially devastating quagmire upon my arrival at the Georgia Department of Education, we learned the following lessons as we worked to straighten out Georgia's testing program:

1. A testing division that contains the appropriate technical expertise and organizational integration and support is foundational to success.
2. Every bit of strategic planning and thought put to use during the development and design of a public procurement is time and effort well spent.
3. Sophisticated procurement solicitations and resulting contracts can only be appropriately developed after having clearly determined *what* to procure. Failure to engage in the analysis needed to make that determination will cost the state in terms of varying cost proposals that may in turn lead to the identification of the wrong vendor partner.
4. One must intentionally fight against the natural influences facing governments in order to build the type of well-crafted RFPs and contracts required for success.
5. There are several useful provisions (payment schedules tied to specific deliverables and a holdback percentage, liquidated damages, and performance bonds) to include within large-scale assessment contracts in order to *encourage* vendors to perform, but each of these protections comes with a financial cost.
6. Strong contract provisions/necessarily are not most helpful in their utilization, but rather in their existence.
7. There is no substitute for an open line of communication where the state and outside vendor can identify and solve the inevitable problems that arise during the life of contracts like these.

Notes

1. I am not aware of any decision on the part of the Georgia Department of Education to pursue litigation against any contractor over the developments mentioned in this chapter. Nonetheless, because the statute of limitations may not have run out on any related claims, the author intentionally takes no position in this chapter regarding the legal responsibility for the damages incurred by the Georgia Department of Education, nor should any of the statements made

herein be utilized to place or remove liability that might otherwise accrue.

2. A performance bond is a third party's agreement to guarantee to the satisfactory completion of a contract. Depending on the terms of the performance bond, they may be utilized to ensure not only quality, but also timely performance of the contract. Performance bonds are sometimes referred to as completion bonds or surety bonds.

3. One approach, as is common practice in some states, is not to allow contract exceptions at all. Arguably one gets a truer cost proposal because bidders are not proposing certain costs based on their prediction regarding their success during contract negotiations.

4. As we previously discussed, a vendor may bid a certain cost banking on the ability to negotiate away some of the more costly protections or liabilities contained in the state's draft contract. This strategy may not become apparent until the contract negotiations break down over a particular exception.

5. It should probably be noted that successful inclusion of many of these contractual protections may depend in large part upon the size and scope of the contract which may in turn be derived from the size of the state or the size of the testing program. Those charged with negotiating contracts on behalf of smaller states that may have relatively smaller contracts at issue may find it more difficult than other states to get the kind of contractual protections I am discussing here.

6. At the time I last dealt with this issue, I believe a more commonly accepted holdback percentage was between 10% and 20%.

12

Project Management and Computer-Based Testing (CBT) Implementation
A Decision-Maker's Guide

Noel Albertson
American Institute of Certified Public Accountants

Introduction

A major trend in test delivery in the last decade is the move from paper-and-pencil testing to computer-based testing (CBT). These transitions are fraught with issues—they can take 2 to 4 years, are often late or over-budget, and sometimes overly optimistic in what they promise to their stakeholders.

This chapter seeks to provide the decision maker with some key practical considerations in organizing a CBT implementation for success around modern project management principles, based in part on experiences in implementing a CBT program for the Uniform CPA Examination® (the CPA Exam). Although formal project management is often viewed as an expensive straitjacket in CBT implementations, if well executed, project management disciplines actually have the opposite effect; the work of content, psychometric, and systems specialists is channeled into productive directions, freeing them to focus on what they do best, saving everyone time, money, and unpleasant surprises.

The information in this chapter may be used for a variety of purposes by those in a decision-making role:

- The executive may use it to help ensure that conditions for success are in place so that he or she feels comfortable delegating the execution of the project and remaining in an oversight role.

- Functional managers may use it to identify any concerns about the productivity and efficacy of their own role and that of their staff.
- External stakeholders may use it to better understand how they can most effectively participate.
- Project managers may use it in working with their management to start a CBT implementation.
- Auditors may use it as a baseline of best practice against which to compare a project that they are auditing.

This chapter is organized into three main sections. The first section, *Why Project Management*, gives some context as to why formal project management has become so central to CBT implementations. The second section, *Organizing the Project for Success*, provides and explains the roles and responsibilities for a model project organizational structure. The third section, *Considerations in Managing the Project*, elaborates on a selection of best practices in managing projects, from the decision maker's perspective.

The Business Scenario Used in This Chapter

In order to be more concrete about the application of best practices we use an example business configuration based largely on the CBT implementation for the CPA Exam, focusing on the project from the exam-owner's perspective, which in the case of the CPA Exam is the American Institute of Certified Public Accountants (AICPA). The example is carried through the rest of this chapter.

In the example business scenario (see Figure 12.1) there are three business entities. The test owner owns the intellectual property of the test and is responsible for developing the test items, producing the exam, and scoring the results. The test delivery provider runs the test centers where the exam is administered and is responsible for distributing the electronic files, administering the exam, and returning the results to the test owner for scoring. The licensing authority is responsible for ensuring candidate eligibility and for score distribution.

The example focuses on the project from the point of view of the test owner, the AICPA.

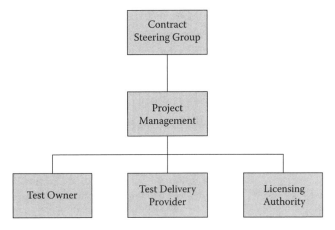

Figure 12.1 Business-level organization chart for a CBT implementation —example.

Why Project Management?

Who Stands for the Whole?

That everyone stands for the whole is an indicator of success in any human endeavor. In any complex project the whole is so vast that one must separate intention from practical reality. One cannot both apply one's specialization and facilitate the application of others' specializations. It is the role of the project manager to take the lead in standing for the whole.

Specialization and Compartmentalization

Specialization requires focus. The tendency toward a compartmental view is an inevitable flip side. Excellence demands specialization, but specialization can lead to suboptimization, by which a part is optimized at the expense of the whole.

As a simple example, suppose a software developer builds a function that enables the user to publish tests for various purposes. From the user point of view, a test published for official administration has to be handled completely differently from one created for a committee review. The former is a production task and may even have regulatory implications, if the testing program is for a licensure exam. The latter is an internal matter better handled by a content expert. If the content expert inadvertently alters an administered test form, the official historical record has been altered. The software developer, however, regards the generalization of the publication function as elegant simplicity, not fully aware that the larger goal has been missed.

An old story also illustrates the point:

> Long ago, a traveler came upon three men working beside the road. He asked the men "What are you doing?" The first man answered, "I am shaping this rock." The second man replied, "I am building a wall." The third man said, "I am building a cathedral."

The first man described an activity—shaping a rock. The second man described a process—building a wall. The third man described an outcome—building a cathedral.

Projects are more successful when team members view themselves as "building a cathedral." For times when the project team loses sight of that vision, the Project Manager is the firewall—the one on the front lines representing the interests of the project's sponsor in the overall success of the project. Whereas others specialize, the project manager stands for the whole.

The Fog of War

As problematic as suboptimization can be, it is easier to manage than some other examples of compartmentalized behavior. All specialists will naturally tend toward remaking the project in their own image, partly out of self-interest, and partly because their worldview is influenced by their profession.

Accomplished professionals want to further their profession. A software developer wants to build *cool* software. A psychometrician wants to rely on modern test theory. A content developer wants richer item types. Hopefully, these aspirations largely align with the business goals for the testing program. But tensions and divergent interests will at times emerge.

When they do, the project manager is the one who owns the success of the whole, the honest broker who facilitates agreements and escalates when agreement is not possible or when the agreed path diverges from the project's charter.

Managing Dependencies

In a complex project such as a CBT implementation, there is as much work in between teams as there is within them. For example, testing programs often implement a diagnostic report that provides the failing candidate with feedback on his or her performance. Anyone who has implemented a diagnostic report knows how much harder this is in practice than in concept.

The technology is simple. Deciding what to report is tricky. A failing candidate wants scores by some subdivision of the exam, as an indicator of what to study. A pass-fail exam does not have the data to provide what the candidates want, but can report some performance short of a set of number-right scores.

If difficulty in producing the diagnostic report is high, the stakes are even higher. All failing candidates receive the report. Frustration can easily mount if candidates disappointed at failing the exam receive a report that they find confusing or unhelpful in understanding why they failed. No amount of psychometric explanations will calm that storm.

Who owns the report? Is it psychometrics, communications, systems, or the content experts? The systems have to provide the data and produce the report. Psychometrics determines what can responsibly be reported. Communications expertise is needed to craft the most candidate-friendly format possible.

Although this is one of the more cross-functional areas, implementing a CBT exam has many more. Without a project manager, those areas that do not fall clearly into one function tend to languish, no matter how important. Each function will do its part and "throw it over the wall."

Meeting the Challenges in Managing CBT Implementation

Formal project management techniques, as distinct from business management, are useful for any project. This discipline, however, is more relevant when a project has certain characteristics. Among these are the project's:

- Size
- Duration
- Complexity
- Number and diversity of disciplines
- Degree to which the work requires cross-functional integration

CBT implementations are moderate in size and complexity, compared, for example, to a large construction project or a government-
funded weapons system. Relative to their size, however, they have an
unusually large number of highly specialized disciplines that must
work in a coordinated fashion in order to be effective.

Making the integration harder, these disciplines have very different education and cultures. Content experts are tied to the subject
matter on which the testing program is based. Psychometricians usually have PhDs and have experience across different testing programs
and subject matters. Systems developers are experts in software
development. Researchers are often systems developers, but usually
inventors who do not gravitate toward production software.

Professional licensure programs often have roots in academia
whose culture of inquiry focuses on exploring the many facets of
an issue as opposed to making a business decision to select a single
course of action. This is well suited for working through the wide-
ranging issues that must be resolved prior to implementation. Once
implementation commences, however, a different, more deadline-
oriented way of managing is needed to move the work forward with
schedule discipline and quick escalation and resolution of issues.

It is at the critical juncture between initial exploration of issues
and implementation that organizing around project-oriented best
practices becomes critical to success. Tasks such as conducting a
practice analysis or deciding on the program's psychometric model
can proceed on a relatively small scale without intensive project
management. Once the implementation starts, the demands of the
scale and pace of the work make project management central.

Organizing the Project for Success

It is surprising how many decision makers fail to make the connection
between difficulties in execution and the root cause of these difficulties in the organization of the project. An illustration of how downstream problems can be tied to project organization is the dilemma of
how to blend existing systems staff with a hired systems integrator.

One way to divide up the systems development work between
staff and the systems integrator is to create two separate systems
project teams with separate project managers and separate systems
architects. Although separate teams may satisfy the preferences of
the staff, such a division of labor has the risk of leading to serious

fragmentation and a lack of an overall design needed to smoothly integrate disparate systems.

The symptoms may manifest themselves in design mismatches. An exam produces results in the test center, but those results are scored in the scoring system. A mismatch is made significantly more likely by having independent systems teams design and build the exam and the scoring system. Better to have separate subteams, but with a single lead architect and a single lead systems project manager. This may be a hard message to deliver to the systems developers on staff, but avoiding it will lead to errors in design that, though they seem like errors of execution, actually have their origins in decisions made on how to organize the project.

There are many valid takes on best practices for organizing projects. What follows is one person's view based on real experience and lessons learned (some of them the hard way) from CBT implementation, and also based upon significant study of the risk factors that lead to project success and failure, particularly in the implementation of information systems. Out of many best practices only a handful were selected because they have been found to address the top tier of risk factors—those factors that, if addressed, tend to result in a positive downstream ripple effect and create the conditions where the project team itself is much more likely to address the secondary risk factors.

Probably the single most important factor in the success of the project, aside from the basic business conditions memorialized in a contract, is having the right team and governance in place, with appropriate, clearly-defined, and understood roles and responsibilities. In this section we explore the basic structure of a test owner's project organization, shown in Figure 12.2.

Sponsor Team

The sponsor team acts as the test owner's primary project-level decision-making body. It typically is composed of:

- Executives of the testing program (executive director and/or vice president)
- A C-level executive, such as CFO, COO, or CEO
- A member of the policy board overseeing the exam
- Other senior decision makers with significant stake in the project outcome

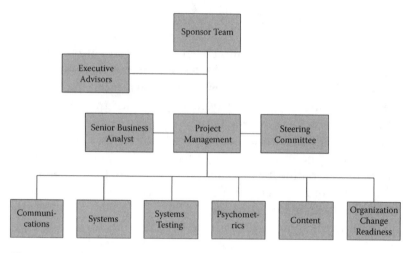

Figure 12.2 Project organization chart for test owner—example.

The sponsor team should be small enough to act as a focused executive decision-making and oversight body. Typically, a sponsor team is not more than six people and ideally only three or four. It is chaired by a person in the role of executive sponsor, who acts as the most senior decision maker as well as the one holding primary accountability.

The sponsor team has the following responsibilities:

- Holds ultimate responsibility for the success of the project
- Reviews and approves project charter
- Ensures funding for the project
- Communicates project charter and emphasizes management sponsorship and commitment to the project's success to the organization
- Sets and manages expectations of and communicates to top management
- Sets standards by which project success will be measured
- Analyzes and makes timely decisions on changes proposed by the project team
- Makes timely decisions to resolve escalated issues
- Reviews and approves project work and progress through key milestones
- Takes initiative to proactively identify and mitigate project risks

The executive sponsor along with his or her supervisor decides on the composition of the sponsor team. The team meets once a month, or twice a month during critical periods on the project.

Executive Advisors

Executive advisors are those senior managers whose expertise and buy-in are needed, but who are not the decision makers for the project. This team often provides a good role for important individuals whose involvement is desired but who are not appropriate for the sponsor team, especially when the executive sponsor is diplomatically attempting to limit the size of the sponsor team.

The executive advisors team has the following responsibilities:

- Provides *subject matter expertise* and *content expertise* input to project requirements.
- Provides ongoing review and feedback through the implementation.
- Provides testing support/validation of newly developed solution.
- Assures communication, support, and buy-in of internal and external customers.

The executive sponsor selects the executive advisors team, based on consultations with a wide variety of internal and external stakeholders. This team meets once a month or as needed.

Project Management

The project management function provides the overall hands-on leadership of the project. This function may consist of one or more project managers depending on the size of the project. For larger projects there are typically several project managers and an overall project director, as well as administrative support.

Much has been written on the responsibilities of the project manager. The following is one experience-based summary of those responsibilities:

- Develops, maintains, and manages execution of a comprehensive project plan that addresses the project's scope, schedule, cost, quality, staffing, communications, risk, and procurement of third-party resources.
- Measures progress against all areas of the comprehensive project plan.
- Defines and executes an integrated change control process.
- Provides a breakdown of the project scope into a series of activities that are sequenced into a schedule; facilitates an estimation

process that produces a realistic overall estimate of project effort, time, and cost.

- Sets acceptance criteria and manages the acceptance process.
- Oversees the creation and execution of a quality plan.
- Manages issue resolution with a particular focus on selective escalation.
- Acquires needed staff and resolves staffing issues.
- Builds team spirit as needed to achieve success.
- Takes initiative to proactively identify and mitigate project risks.
- Reports status to sponsor team, project team, and project stakeholders.
- Works with third-party vendors to define timeline and deliverables and manages them to meet the deadlines and develop quality deliverables.

As the complexity of these responsibilities imply, this is a hard role to fill, especially for niche projects like CBT implementations where the labor pool is quite small. A project manager with a successful track record in delivery of complex, process-intensive large systems is a good bet, as long as this person is partnered with strong individuals in the functional teams and a good business analyst.

Senior Business Analyst

Even more overlooked than the project management role, an analyst with an end-to-end business-level understanding of needed strategic and operational changes is invaluable, and more to the point, essential. Without this role, the CBT processes and supporting systems are often designed by a combination of existing staff (who are steeped in the current way of doing business) and a systems team who are not positioned to understand the long-term business implications of key decisions.

The hard part about *key* decisions is that management often only understands that these decisions were made after they are stuck with them. Such decisions don't present themselves as key at the time when systems and functional teams are wrestling with them. They often seem like arcane technical questions.

An example is the decision to implement Web services. Web services technology enables systems to interconnect and conduct transactions over the Web. These transactions can be very granular. For example, a single test result from a test center can be its own transaction, instead of having a whole day's test results act as one batch transaction, making reconciliation much simpler.

This technology enables a huge advance over traditional batch-oriented processing. The shift, however, from batch to continuous processing has fundamental implications for all areas and levels of the testing program, including what skills are represented on staff, and may necessitate a review of board-level policies. How far back are scores re-reported in case of a scoring error, when each candidate score is its own transaction?

The scope of responsibilities and authority of the senior business analyst role varies widely and can be the subject of passionate debate. The following list indicates some key activities:

- Extracts broad business and strategic imperatives from project sponsors and translates them into the critical implementation decisions for the project.
- Facilitates decision-making sessions with the sponsor team.
- Develops business design and processes to support management direction, based on up-to-date best practices in the testing industry, including the design of CBT systems platforms.
- Facilitates build versus buy decisions for systems.
- Creates business-level requirements for systems development.
- Recommends organizational changes needed to support the new business model.
- Facilitates the deliverable acceptance process.
- Leads business readiness activities such as development of new policies and procedures and staff training.

The senior business analyst should have at least 3 to 5 years of broad management experience in a CBT program, ideally at an executive level, and previous experience with at least one transition to CBT.

Organizational Change Readiness Team

The organizational change readiness team ensures continuity of test development and operations during the transition to CBT. This team works with the senior business analyst to evaluate proposed processes in the to-be CBT organization, and identifies gaps in procedures, organizational structure, and skill sets with the current organization. It proposes and organizes the preparation for the to-be organization.

The goal is that from the first day after CBT launch, the team members can perform their jobs without interruption. They understand their role within the business process, are well trained, have access to the appropriate data, and have the right software and hardware tools to do the job.

Converting to computer-based testing represents a fundamental change in business models. Organizations that do not reorganize their processes and staffing to be ready for such changes are destined to struggle for years after the transition. No amount of blaming staff for nonperformance can remove management's responsibility to manage organizational change before, during, and after the change takes place.

Yet, organizational change is rarely defined as a formal work stream in CBT implementation projects and this is a common problem in information technology projects. The biggest change is the shift from a few events a year to continuous administration. This shift overshadows even the jump in technology intensiveness. Any business moving from batch to continuous processes faces a daunting task of preparation, and a realignment of cost peaks and valleys.

The responsibilities of the organizational change readiness team are:

- To identify human and organizational risks, and to develop and maintain the change charter and organizational change management plan.
- To catalog existing procedures, roles and responsibilities and tools, and to evaluate these against the to-be business model developed by the Senior Business Analyst.
- To identify communication gaps and recommend an approach for addressing these gaps.
- To create plans to implement management decisions on organizational change.
- To lead internal communication and education on organizational change and the team's plan to manage this change.
- To facilitate organizational change events.
- To monitor the organization's receptiveness to the planned changes and to advise management accordingly.

This team is composed of process leads from the other teams and led by someone with broad leadership ability, ideally the senior business analyst. The process leads are those individuals within each functional area with the best overall mix of broad, business-oriented perspective and good process skills.

For a variety of reasons, in practice, managing organizational change tends to remain unaddressed or unfulfilled. One reason is that the organizational change is not viewed as the main event and organizational change therefore does not receive the visible and active sponsorship that managing a change of this magnitude demands. A second reason is that tasks related to change management are often assigned to existing staff members who lack the training for and experience of what is to come.

A third reason is that upper management's attention is focused on what it understandably sees as the main event—the development of the test. A change, however, as significant as a transition to CBT cannot be successfully implemented without change management. And the work of organizational change cannot succeed without visible and active support of the project's sponsors, because much of this work depends on staffing changes that can only be made from the top.

Steering Committee

There is always some inherent tension in any project between the role of the project manager and that of functional managers. It stems from the differing fields of vision—functional managers think in terms of business units, project managers in terms of projects. It is a real tension, based on real choices that businesses make about the balance of authority between the project manager and the functional manager.

If an initiative is clearly separated into multiple projects, the project manager has broad authority to direct the activities of the resources assigned by the project charter. On the other end of the spectrum the project manager role can be more of project coordinator, primarily playing an administrative function. Neither approach is right in all cases. Each project must be addressed on a case-by-case basis.

With regard to CBT implementation projects, one approach that works well is to set up a *steering committee* of functional managers as a tactical decision-making group. This solution adapts well to the sweeping nature of the change inherent in these projects by giving the project manager significant authority, while recognizing and leveraging the important leadership contribution of functional managers. The possible list of responsibilities can vary widely, but, at a minimum, involves the following:

- Review and sign-off on key deliverables where functional acceptance is required (for example, requirements and test plans).
- Address resource issues.
- Periodically meet to review status and provide input.

The steering committee typically meets monthly, or as often as needed.

Systems Team

The mission and composition of the systems team depends on the buy versus build and the in-source versus out-source decisions. These decisions are critical to the success of the project but are outside the scope of this chapter. However, here are a few guidelines that are true no matter which decision is made:

- The biggest factor in staffing the systems team is the scope and complexity of the systems needed. If the exam is multiple-choice only, then using off-the-shelf tools hosted in-house or by a hosting provider should be seriously considered. Another solution to be considered is a complete outsourcing to a service provider such as the test delivery provider. The less distinctive the technology of the exam, the more money and headaches can be saved by outsourcing.
- Hiring a skilled systems team with a proven track record with a similar class of systems is a critical success factor.
- If the systems team is outsourced, making the overall project manager an employee has many advantages. Someone inside the testing program close to the project has to manage the procurement relationship. The overall project manager is the best positioned to do this.
- Building custom software is hard and risky. Using existing tools is preferred unless the testing program has the money to pay for talent needed to build from scratch. Because CBT is a niche market, there is usually not a large enough customer base to support many vendors of off-the-shelf products. So, though custom development has challenges, there is not much of an alternative in the CBT space.
- One key sign of a good systems organization is they use a standard proven methodology for their work. Staff should be trained in the methodology and should have templates and electronic reference versions of the methodology available.

Systems Testing Team

The systems testing team tests the system against the agreed requirements to ensure those requirements are met. This team usually consists of experts in software testing, often supplemented by subject matter experts from functional teams.

Systems testing is a process of greater and greater integration. The first testing performed is at the smallest unit by the person responsible for that unit. Ultimately, the full system must be tested in a series of beginning to end, real-life scenarios with data that is as close to live as possible. For example, the test campaign may culminate in a pilot test in which paid or volunteer individuals who fit the profile of the candidate population apply for and take the test, and receive a score.

The further along the testing, the greater the need for involvement of a broad set of disciplines and views. Initially the team may consist primarily of experts in systems testing. Later, content experts, psychometricians, and other users of the systems should be actively involved in creating and executing test cases.

Systems testing is often the most neglected systems development activity. In the example scenario provided earlier there are three different parties (test owner, test delivery provider, licensing authority) who must integrate their software and test together. A well-run development effort allows as much time for the full range of testing (from the smallest units through to the pilot) as for the development of the software. Business executives and functional managers alike are often surprised at the amount of calendar time needed to ensure that CBT implementation performs smoothly once the program is launched.

The systems testing team is managed separately from the systems team. This does not remove the responsibility for the systems team to perform its own internal testing to ensure that the system is stable enough for systems testing to begin. It does provide for an independent verification against the approved set of requirements.

Considerations in Managing the Project

Although many thorny problems in managing the project have their origin in how the project is organized, a well-organized project still has project management challenges. We describe some critical factors for the decision maker to monitor as evidence of a well-run project.

Make Progress Measurable

How does the decision maker know that reported progress represents true progress? Tracking and reporting progress against plan
is one of the most critical roles of the project manager. And yet the
most common format for reporting status (key accomplishments this
week, plans for next week, issues) lacks the rigor needed to provide
decision makers with the management information that will enable
them to oversee the project.

An approach to status reporting that provides greater transparency to senior management as well as the project team starts with
how the project plan is defined. Status reporting can only be measurable if the plan against which the status is reported is to measurable.
The most practical way to make a plan measurable is to build the
plan around a series of measurable project milestones every 4 to 8
weeks. To ensure that the milestones accurately reflect true progress,
the following guidelines must be observed:

- A truly measurable milestone is deliverable-based. Each milestone is considered complete when one or more project artifacts
  are completed, submitted, and accepted by the sponsor.
- A formal acceptance process provides the best indication of
  whether the team is ready to move to the next step. The features
  of a formal acceptance process are that it is based on preagreed
  acceptance criteria, and results in sponsor sign-off based on recommendations from key management and subject matter experts.
  Examples of deliverables are the content specification outlines
  (CSOs) and the systems requirements document.
- The milestones and their corresponding deliverables should not
  be make-work. They should organically consist of the real artifacts of the project. The 4 to 8 week interval is not arbitrary. If a
  project doesn't produce some concrete finished product regularly
  within that interval, then it is probably adrift. It is important to
  avoid making too many milestones, especially in the middle of
  the project. In the beginning and the end of the implementation
  phase, the deliverables will come more frequently. In the middle,
  however, when the team is doing the bulk of the work, milestones
  between events may stretch out to as much as 12 weeks.
- In selecting the order of what deliverables to produce when, focus
  more on risk than on a conceptual notion of a linear process. In
  the 1970s so-called *waterfall* methodologies for systems development were all the rage. Requirements had to be finished before any

concrete design work could begin. No software could be written before the design was complete. Now we know better. Businesses quickly realized that reality did not work that way. One could not fully understand requirements without engaging in some design and software development. The more unknowns there are, the greater the need for a nonlinear process. If the exam includes complex item types whose scorability is unknown, it is useful to develop a prototype at first. This may involve requirements definition, design, and software development activities prior to signing off for the full requirements for other areas of the program.

With measurable milestones in place, the project manager can report progress against their completion. This will act as an agreed-upon concrete yardstick, and provides the governance needed to build the confidence of stakeholders that the project is on track.

Establish Change Controls

Controls are predicated on the discipline that a project can only have one plan at any one time. The active plan is called the baseline. Controls are intended to ensure that the project executes according to that plan, and that any changes to the plan are formally approved at the right level of management with the appropriate input. Upon approval, this new, modified plan becomes the baseline. This process is known as change control.

The change control process governs how the project handles changes. It typically consists of a procedure, a change request form, and a change control board (CCB). *Change control board* is a generic term for the organizational structure that manages change. In many cases, it is the Sponsor Team.

The first step in the change control process is to determine whether the change rises to the level of requiring the CCB to authorize it. For example, suppose that a psychometrician asks for the design of the screen to be modified so that the item response theory (IRT) statistics are more prominently displayed than the classical statistics. As long as this change does not impact the overall budget and schedule, the project manager may be given the authority to make the decision without having to obtain the approval of the CCB.

For an example of a situation that might require CCB approval, suppose that the test development manager wants to add an approval

step before an item transitions to an operational status. This requirement changes the workflow management for the item bank, which
could be a significant change in scope. This would require a change
request to be submitted to the CCB, which would assess the impact of
the change on schedule, cost, and risk. The CCB would consider the
benefits versus the impact and either authorize or deny the change.

If the CCB authorizes the change, then the project management
team updates the plan accordingly. This plan becomes the new baseline. If the board denies the request, then the current plan remains
the baseline.

In the case of multiparty projects, it is common to have separate
internal and external CCBs. The composition and process of the
external board is typically governed contractually, and not surprisingly there is a large overlap with the composition of the internal
board. With two boards, a change request is approved first by the
internal board prior to submission to the external board.

There are six commonly accepted categories of control: scope,
budget, schedule, risk, quality, and contractual. Each category is
controlled within the overall change control discipline, but may have
its own distinct review process prior to consideration by the CCB.

Manage Risk Proactively

Many common risk factors, such as the lack of sponsorship, have been
addressed earlier in this chapter. Organizing the project well is a good
start on keeping risks to a manageable level. During the execution
phase, formal risk management can help keep risks under control.

Risk management methodologies vary, but most contain the same
basics. Risks are identified, exposure is quantified, and mitigation
plans and contingencies are defined.

Identifying risks is challenging. Many are obvious. Still others are
hidden in plain sight—made invisible by the same blind spots that
created the risks. Others fall into the category of not knowing what is
not known. The first category of risks (the obvious ones) can be identified by management. A combination of outside experts and multidisciplinary internal reviews can identify the second (hidden in
plain sight) and third categories (not knowing what is not known).

Risk exposure is quantified as the product of impact of a risk
times its probability. Rigorous methodologies quantify impact by

calculating the time or money lost. This works well for the biggest risks, where the costs are easier to quantify. For example, if moving the launch date costs $100K/month and the probability of a 3-month delay is likely, then the total risk exposure is $300k × 75% = $225k (see Table 12.1).

If the purpose of quantifying is for a less rigorous purpose, such as ranking and prioritizing risk, then using an ordinal scale may be sufficient. The impact can be measured on a 1 to 5 scale with 5 being catastrophic (project will have to be canceled). Probability can be measured on a scale of 1 to 3 with 3 being likely. The calculation of the risk exposure due to moving the launch date would be 4 (high but not catastrophic impact) × 2 (medium probability) = 8 on a scale of 15. This number is only meaningful in relation to other risks.

A mitigation plan means spending time or money to avoid or lower risk. Sometimes the investment is trivial—the real investment is in the effort to identify the risk early enough to avoid it. For example, if the launch date is very sensitive to the readiness of the production environment, the hardware can be ordered earlier than would be normal. Other times the cost of avoiding a risk is not worth it and management decides to accept the risk and develop a contingency plan to prepare for the possibility.

A contingency is a best-case alternative plan in case a risk cannot be prevented. A contingency plan should include the following elements: the planned action, definitions of the triggering event, clarity on who has the authority to invoke the contingency and the go/no-go date by which the decision to invoke must be made.

An example of risk assessment is shown in Table 12.1. In this table, the risk of a delay is judged to be likely under current circumstances and to have serious impact, but less serious than the potential for gaps in the item bank. By this measurement the latter risk would receive priority attention in resources and focus. Often risk analysis will reveal high impact but low probability risks. These are best dealt with through contingency plans. A classic example of this is disaster recovery.

There are several practical steps one can take to implement a successful risk management methodology.

- Engage one or more external experts with deep experience in large-scale systems projects or with CBT implementations. There is no replacement for the hard-won experience. Projects usually wait until they are in trouble to bring in someone to assess risk. But avoiding problems is much less expensive than fixing them.

TABLE 12.1　Risk Plan Summary—Example

Risk	Probability*	Impact**	Total Exposure	Mitigation / Contingency
#1: Attrition problems in item writing may lead to a shortage of certain topics in the CSOs	2	5	10	Mitigation: Assign key staff to work closely with item writers on these topics. Increase volume of these topics to compensate for potential attrition. Contingency: Create a budget line item to assemble an emergency team of topic experts by a certain date if item levels do not meet targets.
#2: Item presentation issues introduced by the data migration may cause delays or defects in test items to reach the field	3	3.5	10.5	Mitigation: Begin data mitigation design in parallel with CBT systems design. Develop strategy to initiate end-to-end testing at earliest possible date. Contingency: Develop redundant QC late in the test production process. Create a budget line item to assemble an emergency team of temps in case serious problems are identified late.
#3: Uncertainty in systems testing schedule may require launch postponement	2	4	8	Mitigation: Begin integration testing early with visible checkpoints every 4 weeks. Agree on a specific go/no-go date with state boards. Hold a pilot prior to go/no-go decision. Contingency: Agree on a plan for an additional paper and pencil administration with state boards in case of a no-go decision.

* Probability: 3 = Very Likely; 2 = Likely; 1 = Unlikely
** Impact: 5 = May affect validity defensibility of the exam; 4 = May result in potentially damaging publicity; 3 = May negatively impact candidate or state board; 2 = May have significant internal or business partner impact; 1 = Probably has low impact

- Hold periodic full-day multidiscipline review sessions. As expensive as it sounds to pull the project team away from their work for a full day review once every 2 to 3 months, avoiding only one of the major risks that they are likely to find saves more than all the time invested. Risks often emerge from translation problems between different specializations. A content expert says that a certain feature is necessary without realizing how significant the impact is on systems design, even though there are other ways to accomplish the same goal that result in less risk. If a risk review session can uncover this before much energy is invested, then many downstream problems and complexities can be spared.
- Set aside a management reserve of time and budget, held by the appropriate level of management, that can only be released in response to a formal change request (see "Establish Change Controls" section of this chapter for a discussion of how a formal change control process works). This is a seldom-implemented best practice. Executives hate to see what they view as a pile of wasted time or money. But they don't see the hidden costs of padded estimates that project teams use to deal with inevitable uncertainties. By taking this action, executives send the signal that they expect from their project team a true estimate of what the job will take, while holding for themselves the decision of whether and when to allocate all or part of the reserve.
- Set an alternative launch date and a go/no-go date for when to confirm it. In launching a CBT program, moving a launch date is often a complicated issue, fraught with public and even regulatory issues. A paper-and-pencil exam may be administered only a few times a year. To delay at all may mean a 6-month to 1-year delay. Those who administer the exam need long lead times to prepare for the administration, especially for the expense and for securing a site. Setting the go/no-go date provides transparency as to how far in advance a notification of a change is required.
- Hold monthly management review sessions. Projects that hold regular management reviews are much less likely to get off track.

Manage Change

As previously discussed (see "Organizational Change Readiness Team" section), actively managing change is critical to project success. Putting the right team in place was reviewed in our previous discussion. From here, we review the ingredients of a successful change campaign.

The deliverables of an organizational change campaign are:

- A catalog of existing procedures, roles and responsibilities, and tools.
- A process diagram and a set of procedures for the to-be process.
- A transition plan for implementing needed changes.
- A set of training courses, infrastructure investments, and security changes needed to address identified gaps.
- A list of power users.
- A communications strategy and plan, and a large volume of internal and external communication about the changes and how they affect staff and stakeholders.

The purpose of the catalog of what exists is to unearth how work currently gets done. There are typically more surprises here than is expected. People are remarkably ingenious and adaptable in finding ways to get around current constraints, and this knowledge is a gold mine of realism about what needs to be addressed in order for the staff to be productive after the conversion to CBT. At the same time, this catalog will also reveal a myopia that short-circuits important business goals—this too is very instructive in understanding what needs to change. Change is about training, communication, and leadership where current staff can make the leap, staff changes where they are not able, and the wisdom to know the difference.

The process diagram is a deliverable of the senior business analyst. The corresponding procedures are written by the staff members who will be executing them, after enough of the new system is built for them to understand what the new procedures need to be. There is an important synergy here: The same staff members who are writing procedures can and should also be writing test cases and executing them as part of the systems test team.

The senior business analyst plays a key role of ensuring that the change initiative stays on track—that the detailed procedures faithfully implement the to-be process model as well as the business imperatives from the project sponsors.

The transition plan starts with an impact assessment identifying the gaps between the current organization and what is needed to support the to-be process. These gaps are across people, process, and technology categories. The transition plan lays out what steps need to be taken to make the change go as smoothly as possible.

One sensitive issue in any transition plan is staffing changes. There are different philosophies in how to approach such a sensitive topic. One view is to address staff issues in a less public forum and focus the transition plan on training and infrastructure changes. The other is the open book approach—to be clear and transparent about the new roles and responsibilities and the staff transition strategy. Arguing in favor of the latter is that staff members are usually perceptive enough to figure out one way or another that there will be staff changes. By being proactive, management increases the chances that they will retain the most critical staff, and can leverage the staff in making the transition work rather than being passive observers.

Training courses center on the new tools and procedures that the staff must master. They become particularly important where the changing procedures are accompanied by a marked conceptual shift. This is evident in the publishing process. Publishing a test a few times a year is like publishing a book in that what becomes important is the finished product, not the original inputs. *Bluelining* is a common practice in publishing a paper-and-pencil exam. The test form is refined until it is ready to go to print. When publishing is more or less continual, as it is in CBT, it is not as easy to get away with fixing upstream problems downstream. The production team can easily become overtaxed with corrective quality control if the test content is not correct and properly formatted. Even if new procedures and controls address this issue, staff will unwittingly thwart the intended outcome of these procedures and controls if their mental model remains in the pencil-and paper-world.

Infrastructure investments can range from purchasing new PCs for the staff to setting up the production operational environment. The latter often involves contracting with a hosting provider, because the level of maintenance and security required for a CBT operation is often beyond the capability of testing organizations.

The biggest change in security needs in the transition to CBT is that a larger number of people need much greater access to data. It was easier to keep the data under lock and key in paper-and-pencil testing. Greater access needs raises the stakes to a whole new level, typically requiring outside security expertise, or, as previously mentioned, outsourcing.

Power users are the experts in key process areas such as item development, exam production, or scoring. They have both process and system user expertise. They are pivotal players in successfully managing change. Learning a new way of doing business does not take place in a one-time training event. It is a long-term transformation that has to be supported. A critical part of that support is the peer-to-peer help that someone embedded in the organization can provide as part of the day-to-day work.

One successful strategy for developing power users is to take the current process leads and involve them in requirements gathering, developing and executing systems test cases, and in procedure writing. In so doing, they become steeped in the new methods before CBT ever goes live.

Communications strategy has to come from the lead executive of the program and be continual from beginning of the project through the first year of operation. It is an indispensable component of managing the wrenching changes that computerization brings.

Case Study

Through this chapter we explored how the application of formal, state-of-the-art project management expertise to CBT implementation can create the winning situation for all stakeholders. An instructive case example is the work of the AICPA, which, with two other partners, computerized the CPA Exam. We look at a few examples of how the AICPA applied the principles in this chapter in order to draw some lessons learned.

The AICPA's role was that of the test owner. By 2002 it had laid a strong foundation for computerizing the exam. It had done much of the necessary research, had completed a successful practice analysis, had circulated exposure drafts, and solicited comments on the structure and goals about the new testing program. It also was in the process of putting in place a contractual governance structure with a three-way steering committee.

What the AICPA lacked was an internal structure based on projects, organized around the principles covered in this chapter. This became apparent in an independent information technology project risk assessment that the executive director organized, which was a positive step that the program took to lay the foundation for the transition from research to implementation. Fortunately, the program implemented the findings of the assessment, and the changes were made early enough in the project to make a significant impact.

Two years later the project successfully launched on time and on budget.

The CBT implementation was a three-way partnership with Thomson Prometric and the National Association of State Boards of Accountancy (NASBA). What the AICPA lacked internally, the enterprisewide project also lacked. The AICPA's risk assessment recommended the appointment of an enterprise project manager (EPM), reasoning that the CBT implementation faced the same need to create a project infrastructure at the level of the business partnerships. The contractual steering group soon hired an EPM and had him report directly to the chair of the steering group. Although this role was not specifically discussed in this chapter, it is an example of the project manager role applied to the business level. The role turned out to be crucial in that it freed each party from the burden of having to be the overall leader of the coordinated work.

As the AICPA reorganized the project to implement the findings of the risk assessment, it created a steering group and a baseline plan with measurable deliverables. The measurability was particularly important in the contractual relationship with the systems integrator that the AICPA hired to do the bulk of its systems development work. The statement of work tied payments directly to deliverables that had to be approved through a formal acceptance process. Each deliverable was assigned a team of subject matter experts who worked with the systems integrator and then recommended acceptance to the management team. Changes to the defined scope, schedule, and budget were handled through a formal change control process and were adjudicated by a Change Control Board. The discipline of this process helped to create a smooth working relationship and resulted in a predictable process and a set of systems that met the AICPA's needs.

One area where the AICPA struggled was in how to balance the authority of the functional managers (principally the content and psychometric departments) with the project manager. The role of the steering committee (described earlier) has been created on subsequent projects at the AICPA to provide a formal decision-making role for these functional managers. Lacking this clarity in the CBT implementation project meant that issues were often escalated to the executive director who then had to play an arbiter role. Many of these issues could have been resolved without escalation.

Conclusions and Lessons Learned

Each project has its own unique circumstances. The principles discussed in this chapter have to be customized to address those circumstances, resulting in processes and organization structures that may depart from those discussed here. The process of carefully applying them brings the best of what the accumulated wisdom of the project management discipline has brought to the business of projects.

Part V

Monitoring and Improvement

13

Process Management in the Testing Industry

Rohit Ramaswamy
Service Design Solutions, Inc.

Introduction

Process management refers to the set of techniques used to measure and monitor the performance of processes. Process management is a key activity in an organization's quest to continually improve the quality of its services. In the testing industry, as the pressure increases to develop more tests more rapidly, and as the platforms and processes change when developing computer-based tests (CBTs), it becomes more and more necessary to make sure that the processes by which tests are developed, administered, and scored are stable, robust, and consistently produce error-free outputs. In order to ensure this, it is necessary to institute procedures that monitor the key testing processes regularly and that identify points in the process that are especially vulnerable to errors. These points are then potential targets for process improvement efforts.

The need to continually improve requires us to view our work as *processes*, or as sequences of activities for which the entire organization is responsible and to manage these processes. This is opposed to a *functional* view where each activity is managed separately. This chapter presents the key concepts of process management and describes how the methodology can be used in the testing industry using hypothetical but realistic data.

We introduce the importance of the process view by describing the functional view of an organization and its shortcomings.

Functional View of Testing Activities

In the testing industry, as everywhere else, we tend to think of the work that we do as individual tasks. Depending on our role in the testing organization, we may be creating or editing items, assembling items into paper or computer test forms, packaging tests for shipment to a test center, loading tests on to a Web site for administration, delivering tests, or scoring completed tests and producing reports. Because these functions are often the responsibility of different departments in an organization, it is easy to think of these as isolated activities, and to manage and control them as such. Many organizations are set up to manage departmentally, with the staff in each department responsible only for the activities that take place within their own walls. Work that is completed is thrown *over the wall* to the next department downstream. Similarly, work is received over the wall from the department upstream. This *walled* approach, where each department takes responsibility only for its own activities, is called the functional view. A diagram of such a view is shown in Figure 13.1.

What is the problem with the functional view depicted in Figure 13.1? One might argue that as long as each department is ensuring that its output is of adequate quality, it shouldn't matter how the work activities are managed. All that is required is for each department to make sure that it is doing the best it can.

Although this may be true in theory, it never works in practice. There are several reasons for this. The first is that the functional view does not encourage *joint ownership* of the quality of the output. As a

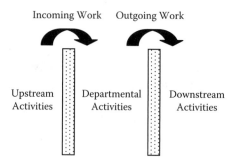

Figure 13.1 Functional view of an organization.

result, there is no common pressure to improve the overall quality of work. As long as each department manager feels that his or her organization is performing to standards, there is no incentive to do better. The second reason is that the functional view is *not focused on the customer of the test*. The customer of the test is not interested in the internal organization of the testing company. If there are errors in the test or in the scoring, or if aspects of the test design or administration are unsatisfactory, customers are not going to express dissatisfaction with a particular department or work group—they are going to blame the entire organization. The department, however, that communicates with the disgruntled customer may not communicate this dissatisfaction to the other organizations, and so they never get fixed. Finally, a functional view *does not provide a structure to manage the interfaces* between activities. As shown in Figure 13.1, the wall represents the interface between activities, and work is received over the wall and is sent back over the wall. The wall is a *gray zone* or a *no-man's land*, and in a functional organization, no one is responsible for this zone. Yet many errors and inefficiencies take place in these interfaces, especially if they represent transitions from manual to automated activities. Units of work (items, test sections, or tests) often pile up at these interfaces and if work triggers and communications between activities are weak, then delays and backups occur. Without proper management of the interfaces, errors and exceptions may slip through these interfaces unnoticed, and may be propagated through the sequence of downstream activities all the way up to the customer.

In summary, the functional view of the activities required for creating, administering, scoring, and reporting tests may result in:

• Customer dissatisfaction, leading to brand and litigation risks
• Errors that remain undetected until they are visible to customers
• Inefficiencies, especially at interfaces
• Little incentive to increase quality
• Poor teamwork, culture of blame and finger pointing, and low employee motivation

Given these limitations, a functional view is not viable in the current testing environment if a testing organization is to remain competitive, maintain its brand and reputation, and avoid the risk of litigation. A process view allows for customer focused, end-to-end management of all activities involved in testing. We now describe the elements of a process view.

Process View of Testing Activities

The process view takes a systems approach to testing activities. Rather than focusing on individual organizations or departments, the emphasis is on the sequence of activities needed to successfully create, administer, score, and deliver reports to the customers of the test. Success is jointly determined by the business and the customers, and metrics are used to determine if the processes are successful. Each activity in the process adds value to the customer or to the business, and the management of each step in the process is undertaken with the conscious understanding of each activity's impact upstream and downstream. Work in an organization is divided into work processes and one process feeds the next. A high-level process view (called a Level 2 map; this terminology is explained later in this chapter) of creating a test is illustrated in Figure 13.2. As can be seen from this figure, the walls from Figure 13.1 have vanished. There are still discrete steps that indicate the different activities that need to be performed, but the arrows represent the sequence of activities and visually indicate that all activities need to be completed in order for the test creation process to be complete. The recipient of the output of the test creation process, whether paper-and-pencil or CBT, is the *customer* of the process.

In the process view, the notion of the customer extends beyond the output of the process. In Figure 13.2, each following step is the customer of the preceding step. For example, those who create the test in Step 5 of Figure 13.2 cannot do so without reviewed and revised items from Step 4; they are the customers of the item review process. This *customer–supplier relationship model* is an important component of the process view of organizational work. By treating the next step as

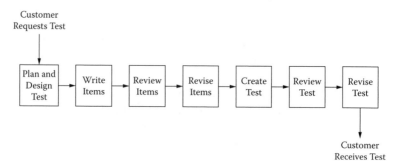

Figure 13.2 The test creation process—Level 2 map.

the customer, a process view encourages each step in the process to pay attention to meeting the requirements of its downstream customers. Downstream steps are encouraged to provide feedback. Therefore, the process view encourages attention to quality and good communication, which does not take place in the functional view.

When the process is mapped to lower levels of detail, the flow of information at the interfaces can be recorded so that the management activities that need to be performed can be identified. In addition, at lower levels of detail, individuals or organizations responsible for each process step are indicated in the process flows so the interaction between the individuals responsible for various parts of the process can be clearly identified.

As a result of the process orientation, the organization can have much greater confidence that it can create and sustain tests that are effective, efficient, and meet the needs of both the test takers and the institutions requiring the tests. Although a process view is a necessary criterion for success, it is by no means a sufficient criterion. The processes need to be managed continuously and diligently to ensure that they are performing as required. The activities required to do this are referred to as the process management system. The rest of this chapter is devoted to explaining the elements of such a system.

Where Do We Begin?

There is a chicken-and-egg problem as we think about viewing testing activities in terms of processes. How can we manage processes continuously if the processes do not exist in the first place? A prerequisite to process management is that processes must be mapped so that a pictorial view of the process activities is available. From the process management perspective, it is adequate as a starting point to map the process in its current operating form—this is called the *as-is* process. The as-is process is often mapped by a team of subject matter experts who are responsible for the process tasks. The advantage of this approach is that it provides a quick starting point for developing measures to manage the process that will help to identify areas that need improvement. The disadvantage is that the as-is representation may not be an accurate depiction of the process, and that the current measures of performance may not reflect the real process performance, because work may be done in different ways by different individuals. For processes that are particularly risky, a better approach is to design a process from scratch, and ensure that the process is designed to effectively meet the needs of its customers. This systematic approach is called *process design* and is a precursor to process management. The disadvantage of doing a formal process design before engaging in process management is that the new process needs to be implemented, tested, and stabilized and this is a substantial investment of time and resources. Clearly, an organization cannot design all its processes from scratch at the same time.

There is no formula for deciding whether to begin with a formal process design or whether to begin with monitoring a quickly mapped as-is process. The approach depends on the resources available, the risk involved, and the time frame for implementation. Whatever approach an organization begins with, it is important to realize that process management and design are part of a continuum, and any selected approach should be considered a starting point to get to the other approach as necessary.

Irrespective of the path chosen to get to process management, the point to remember is that process management is an *ongoing, continual improvement activity*, whereas process design is a *discrete* activity. An organization may choose to take some time to design processes, but this is not an end point in itself. The long-term success and effectiveness of testing processes depends on how well they are managed over time.

Roadmap for Implementing a Process Management System

Organizations seeking to implement a process management system for testing processes can follow a roadmap that consists of the following steps:

1. Identify and map organizational processes.
2. Identify the critical output measures for the processes.
3. Identify the process measures that map to the output measures.
4. Develop a plan for monitoring the output and process measures.
5. Develop a plan for analyzing and reporting process performance data.
6. Create a team to review the analysis and reports and take action as necessary.
7. Embody the continuous improvement philosophy.

Each step is described in greater detail in the rest of this chapter.

Step 1: Identifying and Mapping Organizational Processes

The first step in the implementation of process management system is process identification and mapping. Process mapping involves the pictorial depiction of the activities of a process in the form of a flow chart. Clearly, it is possible to depict process activities at various levels of detail; at the highest level, all major activities for an organization can be represented by just a few boxes; at the lowest level, each task can be split into more and more minute subtasks. There is no magic formula that determines the right level at which processes need to be mapped for successful process management; a rule of thumb is that the processes should be mapped to an adequate level of detail that ensures that all key activities are represented, but should not be mapped to so much detail that it becomes difficult to easily acquire an end-to-end view of the process. Each organization must determine its own appropriate level of detail.

The recommended approach to process mapping is to start at the top and work down. The first task is to list all the processes that pertain to the organization. These could be the processes that are directly involved in creating, administering and scoring a test (these are called the *core*, or *operating* processes) as well as those that are needed to support the business of the organization (these are called the *enabling*, or *supporting* processes). As mentioned in chapter 1, for a testing organization, the typical core and enabling processes are the following:

Core Processes
- New Product Development
- Test Creation
- Test Ordering, Delivery, and Administration
- Scoring and Reporting
- Customer Relationship Management and Communication

Enabling Processes
- System Development and Support
- Marketing
- Quality Management
- Security Management
- Billing

This description of the key processes of an organization is called a *Level 1 process map*. It is always useful to start a process management initiative with a Level 1 map because it allows the organization to view the entire body of work that needs to be performed to make the business successful. An organization may decide not to include all the processes in its process management initiative; a typical decision is to focus initially on the core processes. But even in this case, outlining the Level 1 process map that outlines all processes is a good practice because it effectively describes the entire organization's scope of work.

The Level 1 process map is usually merely a description of an organization's range of work activities. To understand what work actually gets done in an organization, the core processes and some critical enabling processes need to be mapped to the next level of detail, called the *Level 2 process map*. For each Level 1 process, this map describes the 5 to 7 steps that represent the most important activities. As mentioned earlier, Figure 13.2 is a Level 2 map of the test creation process.

The Level 2 map provides a good descriptive overview of process activities. But in order to improve processes, organizations often find that the Level 2 map is not detailed enough to get a deep enough understanding of the process activities. The next level of detail, the Level 3 process map, expands each step of the Level 2 map to 5–7 steps, and adds decision points and organizations responsible for the various activities. A Level 3 process map of the *review and revise* step of the test creation process in Figure 13.2 is shown in Figure 13.3. The process map in Figure 13.3 describes process activities in terms of the organizations responsible for those activities (e.g., external subject matter experts, test development, and the item banking group). Such a map is also called a *swim lane chart* or a *deployment flow chart*.

Most organizations implementing process management document their processes at least up to the level of the swim lane chart.

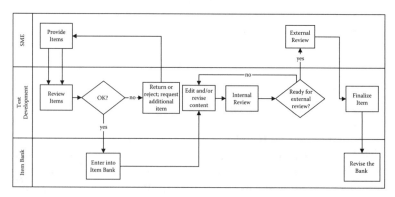

Figure 13.3 Review and revise step—Level 3 map.

Lower levels of detail (Levels 4, 5, and 6) are possible, and are used by organizations for training documents and for documenting detailed procedures, but it is usually not necessary to map processes at this level of detail for process management activities.

Step 2: Identifying Critical Output Measures

The output measures are indicators of the quality of the output of a process. These are usually the results of process performance that are visible to the customers, affect the financial bottom line, or impact the reputation or brand of the organization. Some typical examples of output measures that might apply to testing processes are:

- *Financial*—Revenue, operating expenses, margins, market share
- *Brand Related*—Customer satisfaction, intent to purchase additional tests from the company, likelihood of referral of testing company to other sponsors, customer complaints
- *Performance Related*—Cycle time for new test development, number of error-related incidents, productivity of staff, system uptime for CBT and so forth.

Most organizations begin with common measures such as customer satisfaction, cycle time, errors, and cost as the output measures for process management, and add other measures as their process management systems mature. There is no rule that these are the only measures that should be selected at the start; the organization just needs to be careful that the effort required to collect data and report on multiple measures does not dilute the focus of its process management effort. For most testing organizations, errors in tests are sources of great risk and have the potential for serious financial impact on the organizations. If only two measures can be initially chosen for implementation, then errors and customer satisfaction should be these two.

Step 3: Identifying Process Measures That Map to the Output Measures

Process measures, also called *in-process metrics*, are indicators of how the various process steps perform. These measures, defined at each step of the process, assess the effectiveness and efficiency of the activities that need to be performed to produce an acceptable process output.

Effectiveness measures reflect the quality of process output; efficiency measures reflect the cost incurred in producing the output. The process metrics are called *leading indicators* of output performance because deterioration of performance within the process could be a precursor to deteriorating performance at the output of the process, which may be visible to the customer and have financial implications. We need to select the process level measures that must be monitored regularly in order to ensure that there is no performance deterioration in the outputs.

Examples of effectiveness and efficiency measures in testing processes are:

Effectiveness Measures
- Cycle time for item preparation
- Cycle time for internal item review
- Cycle time for external item review
- Cycle time for test assembly
- Cycle time for scoring tests
- Number of errors (by type) created in item preparation
- Number of errors (by type) detected and corrected in item preparation
- Number of steps between error creation and error detection

Efficiency Measures
- Number of hours of work needed to get an item into the item bank
- Number of hours of work required for internal item review
- Number of quality control (QC) inspections in the development process
- Number of hours spent in rework to correct errors
- Number of steps that need to be reworked to correct reworks

The metrics are intended to be general examples, and do not pertain to any specific testing organization—each organization needs to design its own set of relevant metrics.

A large number of process metrics can be created for each process step. From a process management perspective, we just need to concentrate on a few, critical process metrics. How do we identify these metrics? Clearly, the most important process metrics are those that are leading indicators of the output metrics selected in Step 2. For example, if the number of errors that lead to incidents is an output metric of concern, then the number of errors created at each process step is a process metric of interest. If the cost of producing or administering tests is the output metric that the organization is

TABLE 13.1   Mapping Process Metrics to Output Metrics

	Process Metric 1	Process Metric 2	Process Metric 3	Process Metric 4	Process Metric 5
Output Metric 1	X				
Output Metric 2		X		X	
Output Metric 3					X

focusing on, then the resources used and the amount of rework are process metrics that should be monitored. Finally, if cycle time to produce tests is an output measure of concern, we should be focusing on the time each process step takes for completion.

There is no rule for determining the appropriate process measures. A common method is to brainstorm all possible process measures and to place them in the columns of a matrix. The output metrics are placed in the rows of the matrix. The process management team reviews all the metrics in the columns and maps them to the metrics in the rows. An example is shown in Table 13.1.

In Table 13.1, Process Metric 3 does not map to any of the output metrics and so is not likely to be a candidate for ongoing monitoring. Even after the mapping exercise shown in Table 13.1 is completed, there may be too many process metrics to effectively monitor. If this is the case, the metrics must be prioritized to identify those that are most important. For example, if we are monitoring errors in process steps, there may be some automated steps where the number of errors is typically very small. Or if we are monitoring rework, there may be some error correction steps (such as correcting typos) that are instantaneous, and monitoring the time taken to correct these may not be particularly important. The objective of identifying process metrics should be to come up with a set of 7 to 12 metrics across all processes that are strongly aligned with the output metrics of interest. These are the core metrics of our process management system.

Step 4: Developing a Monitoring Plan

For many process management applications, defining the metrics is the easy part. The more difficult aspect is to be able to regularly collect data on the key metrics that have been defined. Yet the development of a monitoring plan is the most critical step in a process management implementation; if we are not collecting data that is timely or accurate, the whole value of the process management program is undermined. It is therefore critically important to develop a monitoring plan that will determine when, how, and by whom the data will be collected. The main components of a monitoring plan are:

- Operational definitions for the process and output metrics
- Sampling schemes and frequency
- Data collection methods
- Storage and access
- Quality control

Developing operational definitions for metrics is a critical activity that is often ignored. *Operational definitions* refer to clear and detailed rules about how a metric is calculated, and what is included and excluded from the calculation. For example, consider the definition of the number of errors at a process step. Although the definition seems obvious at first blush, there are many different ways in which it can be defined. For example, are we actually referring to the average number of errors over a time period? Or would we be more interested in the maximum number of errors? And what is an error anyway? Is it something as basic as a typo that is corrected instantaneously, or is it something bigger? And how do we define *bigger*? Is it something that is only detected at a downstream step? But what happens if an error is discovered in the step in which it is made, but it takes time to correct? Clearly, differences in operating definitions influence how the process performance gets measured and analyzed, and inconsistencies in the definitions lead to inconsistencies in results. It is therefore important to ensure that realistic, logically consistent definitions for the output and process metrics be developed before any data gets collected.

Sampling is another important consideration. How often should data be collected at each step of the process? In the absence of automated data collection systems, the collection of data becomes a time-consuming activity, and it is important to ensure that data collection effort does not get in the way of actually doing the work. A carefully planned, unbiased sampling scheme needs to be designed to collect data without unduly taxing the system. The sampling scheme is connected with the *data collection method*. If data is collected manually, it is important to ensure that user-friendly check sheets and templates are available that can be completed without a lot of effort. These must be reviewed for error and bias and the data collectors must be trained to make sure that the data is collected accurately and consistently. It is also important to identify the members of the organization who are going to be involved in data collection. In some cases, it may be appropriate to have the data collected by those who are working on the process steps—this is efficient, but may give biased results. Appointing quality analysts who are external to the process to collect the data may result in objectivity, but may also result in organizational resistance at the outsiders who are collecting data. The organization needs to balance bias considerations with efficiency. If the data is collected manually using templates and check sheets, the question of storage becomes important. The data must be entered and stored in a system where it can be easily retrieved and analyzed. It is important to ensure that the data entry is done promptly, and that error checks are conducted regularly. Over a period of time, as the process management system matures, the design of a robust database becomes important and issues related to permissions for access need to be considered. It is good practice to think about database design issues early in the process management initiative. Finally, we need to make sure we implement *process management about process management*. Storing, managing, analyzing, and reporting on process performance is a process in itself, and we need to make sure that this process is managed effectively over time. *Quality control activities* that regularly assess data quality, currency and integrity, the accuracy of automated calculation modules, the efficacy of the sampling scheme, and the consistency of any manual data collection efforts are very important to maintain the integrity of the process management effort.

Step 5: Developing an Analysis and Reporting Plan

Analysis and reporting of process performance data is a critical component of process management efforts. Without effective analysis and reporting, a process management program cannot be successful. Careful thought must therefore go into the design of the kinds of analysis that need to be done, and the way in which these analyses are displayed in process management reports. Typically, a process management report consists of three parts:

1. A section describing the output metrics, usually over time.
2. A section describing the process metrics, by process step or over time.
3. A section describing the inputs, such as volumes or resources.

Output metrics are usually represented as trends over time or as *defects* where the performance did not meet some customer standard. The reason for representing these metrics in defect form is that they measure process performance on aspects that are visible to the customer (e.g., on timeliness, or on errors detected by customers). Poor performance on these aspects can result in significant customer dissatisfaction.

Displaying the extent to which the process does not meet customers' requirements helps to focus improvement efforts. An example of how output metrics are displayed is shown in Figure 13.4. *Note that these data are for illustration only—they do not represent any real testing situation.* The metric is the number of errors discovered by subject matter experts (SMEs) or items rejected by SMEs during the review of a test during test creation. This represents Step 6 of the test creation process in Figure 13.2. Figure 13.4 shows data taken over multiple tests. Suppose market research has shown that on a review, customers (SMEs) can tolerate up to 3 errors/rejects per test before they feel that test developers have done a poor job of item review prior to test assembly. The customer tolerance level is shown by the flat line in Figure 13.4.

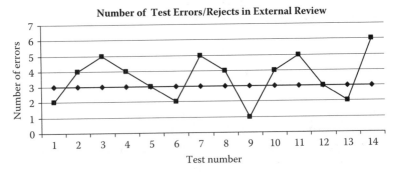

Figure 13.4 Test errors/rejects in review compared to SME requirements.

As Figure 13.4 shows, the actual numbers of errors in the test review step show that there is considerable variability around the desired maximum requirement of 3 errors/rejects. The question is whether this variability occurs because the test creation process itself is flawed, and there are a variety of random causes such as unpredictable typos, key errors, or mistakes in item construction. Or are there some unique, particular causes, called *special causes* (such as inexperienced test developers) that can be identified and addressed? The reason that this question is important is that if the variability is part of the natural variation of the process, then the only way to reduce this variability is by redesigning the process. Because there is no special cause that can be directly identified, the task of process improvement becomes more difficult.

To answer this question, process output metrics can be displayed in the form of *control charts*. Control charts predict the variability that can be expected from inherent randomness in a process as *control limits*. Points that lie within these control limits are part of the normal process variation. Points outside the control limits can be attributed to special causes.

Figure 13.5 shows a control chart for the data shown in Figure 13.4. The lines labeled UCL and LCL represent the upper and lower control limits respectively, which, as mentioned previously, are the

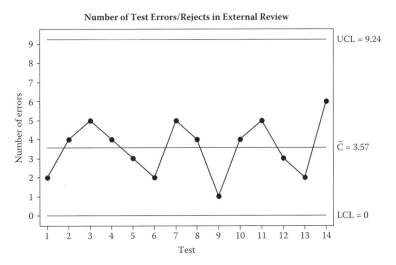

Figure 13.5 Control chart of test errors/rejects in review.

predicted limits for natural process variation. Because all the points are within the control limits, it can be inferred that the test assembly process is under control and that the variation in the average number of errors is normal and expected. That does not mean, however, that the variability is *acceptable to the customer*. This is a very important distinction. A process that is in control is stable in the statistical sense, but its variability could still be unacceptably high. Figure 13.4 shows that for 8 out of the 14 tests, the number of errors is higher than 3. This is clearly unacceptable to the customer. An aggregate metric such as the *% of tests with errors >3* collected over an appropriate time period can provide an indication of the extent to which the process performance meets customer expectations.

In summary, the metrics shown in Figures 13.4, 13.5, and 13.6 are all output metrics and indicate different aspects of process performance. There is no magic formula that indicates which output measure is most appropriate. The output measures that must be selected are the ones that are most relevant and indicative for the particular process and application.

Process metrics are indicators of performance of process steps that contribute to the quality of the process output. Clearly, the number of errors at the output of the test creation process of Figure 13.2 depends on the errors committed within the process. One obvious process metric could therefore be the number of errors committed at each step of the test creation process. This metric will help us identify the *error creation steps* in the process. These are steps to focus on during a process improvement effort. Not all errors, however, that are created during the process are detected by customers. Many of these errors are corrected at various steps of the process. So it would also be useful to identify the steps where most errors are corrected—these would be *error correction steps*, and would indicate steps that may need to be preserved for process quality. Finally, for each process step, it may be useful to know the percentage of errors that go undetected—this gives us an idea of the steps that are high-risk and need to be improved.

Figure 13.6 shows the average number of errors created and corrected at each step of the test creation process shown in Figure 13.2. The average total number of errors created and detected in the process is 36.2. There could be errors that are not detected in the process, but obviously they would not be identified in Figure 13.6. Figure 13.6 shows that the average number of errors detected in the

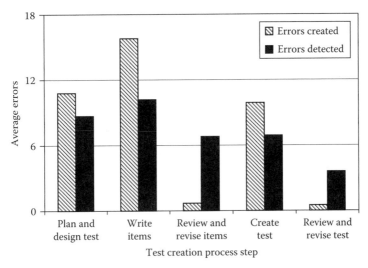

Figure 13.6 Average errors created and corrected by process step.

test review step of the process is about 3.6, which is consistent with the center line of Figure 13.5. Figure 13.6 also shows that errors that are created in the test planning, item creation, and test creation steps are not all detected and corrected at source, and that the review steps play a major role in the correction of errors. Significant opportunities exist for process redesign to eliminate the occurrence of errors.

Finally, *input* metrics, by definition, measure the inputs to a process. The volume or quality of inputs can affect the quality of the process steps and outputs, so tracking inputs can help to uncover causes for poor process performance. Figure 13.7 shows the average number of tests assembled per week for the 10 weeks of our study. The figure shows that there is a fair amount of variability in the number of tests created every week. If the organization is not staffed to handle the weeks when the volume is high, the staff may be overburdened and this may contribute to deterioration in quality.

Analysis and reporting of process results are critical activities for successful process management. They are also, however, the most time-consuming activities, especially if performed manually, and require analytical skills. Organizations that are committed to process management invest in software to automate the analysis and reporting process, so that standard reports are automatically produced and distributed. The data is usually available in databases to

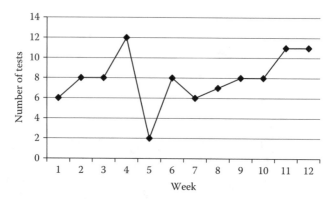

Figure 13.7 Number of tests created per week.

create custom reports and charts. Even if the analysis and reporting is not automated, the process must be designed to be able to produce reports regularly and efficiently. For example, spreadsheet macros can be used to perform repetitive analyses at the touch of a single button. Care should be taken to ensure that the effort required to analyze the data and produce reports is not so overwhelming that it detracts from the primary purpose of the analysis, which is to identify opportunities for process improvement.

Step 6: Create a Team to Review the Analysis and Reports

Reporting on metrics and analyzing process performance is of limited value unless there is a governance structure that is able to review the results and to take action if necessary. Two levels of governance teams are commonly needed to manage results. Each process will have a *process management team*, and the process management team consists of a *process owner* who is responsible for managing the performance of the process. Typically, an organization would have a process owner for each major testing process, for example, item and test creation, test administration, or scoring and reporting. The process owner role is a coordinating and facilitating role—the person who is selected to be the process owner should be accountable for process performance and should be able to influence other members of the team to make changes where needed, but does not need to be the manager or organizational leader of the process group. In fact, it is better if the process owner is not the most senior manager in the organization because effective process management requires the ability to build cross-functional coalitions, and traditional hierarchical leadership often gets in the way. The best process owners are often middle to senior managers in the organization who are able to facilitate and motivate teammates who may be their peers or even their hierarchical superiors.

A *process management team* (PMT) supports the process owner. This team consists of individuals responsible for various steps of the process. For example, in the case of the item and test creation process, the PMT members could be the representatives of the item banking process, the editorial process, the test assembly process, the review process, and so on. The process owner and the PMT should meet at least once a month, and preferably twice a month, to go over the reports and analysis and to identify areas where improvements may be needed. If there are performance areas that need improvement, the process team member responsible for that process step will be charged with creating a team to improve the process.

The last piece of this governance structure is a *process steering committee*, which is composed of the senior executives in the organization, typically responsible for the departments that are in charge of the key process activities. The steering committee is collectively responsible for all the process outputs, as well as for the customer and financial measures relating to these outputs. At least once a quarter, and preferably monthly, the steering committee gets together to review aggregate process results, and the associated customer and financial data. These enterprise-wide aggregated reports are called process dashboards or scorecards, and are often displayed on the walls of organizations as a visible symbol of the organization's quality. New process improvement or process design initiatives may also be launched by the steering committee based on the dashboard results. These initiatives are more likely to be cross-process or enterprise-wide in nature.

Conclusions and Lessons Learned

The process management framework is a powerful practical model for ensuring quality in the testing process. The process view provides the opportunity for end-to-end oversight of a process instead of managing narrowly defined functional areas. The carefully defined measurement structure allows for focused monitoring of process performance in areas that are important to the customer. And finally, the process management team ensures that the results are analyzed, reviewed, and acted upon. In this chapter, the test creation process is used as an example to illustrate the concepts, but the procedure can be used for all core and enabling testing processes.

Although the process management framework is conceptually attractive, its success ultimately depends on the commitment of the organization to make it work. Process management represents a continuous improvement philosophy in action, and without a dedication to this philosophy, no process management infrastructure can sustain and survive. Even the best process management designs can fail due to indifferent senior management, inadequate process documentation, lack of teamwork, inconsistent data collection, and insufficient analysis. If properly applied, process management can be a powerful instrument to create better quality, satisfy customers, and mitigate risk in the testing industry.

14

Six Sigma in Testing

David O. Anderson

Educational Testing Service

Introduction

In recognition of the rapidly changing and increasingly competitive educational testing environment, the Educational Testing Service (ETS) recently instituted a Six Sigma process improvement program. This methodology offers tools to increase performance and to reduce rework, and improvements that ultimately lead to greater efficiency and increased revenue. Simply put, Six Sigma is a disciplined, data-driven approach to eliminating errors and inefficiencies in business by applying proven methods to streamline processes and add increasing value for the customer. It uses data and metrics to reduce costs and improve the quality of output, while meeting deadlines and saving clients money. The approach can be applied to just about any type of business, industry, or activity.

To date, over 80 staff have been trained, and their various projects have produced almost $10 million in documented savings for ETS. Projects have included improved call center labor forecasting and scheduling, significantly shorter item development time, reduction of answer sheet scanning rejects, and better use of test center capacity, among many others.

This chapter provides a description of Six Sigma methodology, an overview of its implementation at ETS, several examples of Six Sigma projects at ETS, and advice for other organizations that might implement Six Sigma.

Introducing Educational Testing Service

For nearly 60 years, the nonprofit Educational Testing Service has developed and administered tests for a variety of purposes around the world. Some of our most recognized tests include the SAT®, TOEFL®, the Advanced Placement Program (AP®), and GRE®. ETS develops, administers, and scores more than 24 million tests annually in more than 180 countries, at over 9,000 locations worldwide. Through the years, ETS has seen the emergence of new technologies and methodologies for the rapid analysis of students' responses and reporting of scores, as well as for the actual administration of test questions. Even with these new technologies, however, the fundamental processes and procedures for handling test registration, question writing and review, printing and distribution of test booklets and answer sheets, customer service, finance, and so forth, have evolved haphazardly over many years and stand to benefit greatly from systematic analysis and elimination of the redundancies and inefficiencies such analysis reveals. With the emphasis that educational testing plays as the result of the No Child Left Behind (NCLB) law and the current competitive environment, every testing company—old and new—is under increasing pressure to perform faster, better, and cheaper. ETS is no exception.

Six Sigma Methodology

Six Sigma is a highly disciplined problem-solving methodology that seeks to identify and eliminate causes of errors, defects, and failures in business processes by focusing on outputs that are critical to the customer. Through the application of statistical methods, Six Sigma seeks to reduce and control variation in process performance, which in turn, improves process efficiency.

Six Sigma originated in manufacturing companies like Motorola (Pyzdek, 1997) and General Electric in the 1970s, "for production processes because of the high volume and high degree of standardization that define such activities. The goal was to eliminate waste by achieving near-perfect results" (Biolos, 2002, p. 3).

Six Sigma offers tools to increase performance and decrease process failures, improvements that ultimately lead to greater efficiency and increased revenue. It is a technical problem-solving methodology, focusing on financially measurable results, and striving to provide customers with the most efficient, cost-effective services that meet their needs.

The Six Sigma approach is based on the belief that the number of *defects* in a process can be systematically measured and prevented, thereby getting as close to doing it right the first time as possible. This is certainly an improvement over fixing the process through rework before it gets out to the client or customer, or fixing the output after it is angrily returned.

The emphasis of Six Sigma methodologies is on decision-making based on facts and data, rather than on assumptions and hunches. How often have you said, "They are always late; why don't they hire more people?" or "If only they had slowed down, they wouldn't make so many errors." Solutions are often offered with little or no pertinent data. Six Sigma requires the collection and analysis of relevant data, together with objective analysis, prior to modifying a process to reduce or eliminate the problem.

The most desirable projects for a Six Sigma program are those that have high business impact (large dollar savings with visible effects), focus on urgent problems, and have a high probability of success within a limited timeframe. A partial list of problems and issues that can occur in testing companies is shown in Table 14.1. Each of these can become the focus of a Six Sigma project.

Key Six Sigma Concepts

Cost of Poor Quality

An important concept in Six Sigma is the cost of poor quality (COPQ); that is, the cost of producing a defective product (e.g., candidate bulletin, test booklet, score report) or service (e.g., call center response, honoraria payment, test center staffing). Beyond the very visible direct costs associated with finding and fixing these types of defects, COPQ also includes hidden costs, such as (a) lost customer loyalty resulting from failing to meet customer expectations the first time; (b) missed opportunities for increased efficiency, (c) increased cycle time, (d) labor and cost associated with ordering replacement materials, and (e) costs associated with disposing of the defective materials. COPQ can also be discussed in terms of excesses leading

TABLE 14.1   Examples of Candidates for Six Sigma Projects in a Testing Company

Customer Service Area
- Variation in handling time by customer service representatives
- Large numbers of repeat calls
- Inefficient workflow for handling customer complaints
- Low up-sell conversion rate
- Poor and incomplete information collection
- Inaccurate call volume forecasting and staff scheduling
- Unacceptably high level of abandoned calls
- Slow response to e-mail inquiries

Test Administration Area
- Inaccurate testing volume projections and test center scheduling
- Expensive variation in candidate bulletin content across testing programs
- Excessive cycle time to process requests for testing accommodations
- Excessive costs for expedited shipping to candidates and/or test centers
- Overstaffing of test centers
- Nonoptimal use of test center space
- Late or incorrect delivery of materials to test centers
- Excessive variation in production cycle times (test booklets, study guides, bulletins, etc.)

Test Development, Scoring and Reporting Areas
- Large proportions of newly written test questions fail to meet psychometric specifications
- Inaccurate forecasting of required numbers of readers at CR/Essay readings
- Late delivery of score reports
- Inaccurate results on candidate score reports or state summary reports
- Excessive time resolving registration and answer sheet scanning problems
- Excessive cycle time for Standard Setting (Cut Score Setting) report production

Finance Area
- Variable profit margins from contracts
- Excessive costs for test administration services
- Unacceptable delays in paying honoraria
- High proportion of errors on honoraria payments
- Excessive number of bad checks received
- Excessive number of customer refund requests
- Excessive cycle times to handle purchase orders
- High volume of credit card declines
- Noncompliance to corporate travel policy

Other Areas
- Perceived poor use of administrative and secretarial support staff
- Excessive inventory costs and inefficient use of warehouse space
- Nonstandard and inefficient proposal writing process
- Nonstandard methods to cost out proposed new work
- Delayed removal and reuse of surplus property from work areas

to waste—overproduction of paper bulletins, extra transportation costs due to expedited delivery, excessive delays, and so forth.

Voice of the Customer

Another important Six Sigma concept is the voice of the customer. It is vitally important to understand that it is your customer who defines quality. You may believe you know what the customer wants, but until you ask, really ask, you will only be guessing or hoping. You need to determine what the customer thinks is important to address, aside from the obvious such as scoring and reporting errors or overdue score reports. Customers may be happy with your published score-reporting schedule, but they expect you to stick to that schedule, no matter what. Test center staff and essay readers expect their honorarium checks on time, no matter what. Customers want predictability and consistency. Inconsistency or variability is unacceptable in today's competitive environment.

Defects per Million Opportunities (DPMO)

Six Sigma strives to ensure customer satisfaction by focusing on delivering products and services so reliably that there would be only a .0003% chance of the customer encountering a problem. This translates to only 3.4 defects per million opportunities (DPMO). Six Sigma aims for 99.99966% accuracy; 99% accuracy is only 3.8 Sigma. For every one million opportunities, this might mean going from 5,000 misplaced answer sheets a year (99% accuracy) to 1 or 2 (99.99966% accuracy), or going from 200,000 misaddressed score reports a year to 68, or even going from 10,000 incorrectly reported test scores a year to 3 or 4. Each of these improvements is highly visible to the customer and would be highly valued for continued and new business.

DMAIC Process

Although there are a number of different ways to implement Six Sigma, the five-step methodology used at ETS is the structured DMAIC approach: define, measure, analyze, improve, and control.

DMAIC is the most popular Six Sigma methodology in use today. The completion of each of these five steps is key to a successful Six Sigma project.

1. **Define** the problem. Identify problems, choose the one that will have most cost-benefit if resolved, identify the process affected by the problem, and then focus on that process.
2. **Measure** the extent of the problem. Collect data, calculate the frequency of defects/errors, and compare that to total opportunities for making errors in the process. This gives your baseline DPMO (defects per million opportunities), against which you can measure your progress. Identify where in the process the majority of defects and errors are occurring.
3. **Analyze** the data to determine root causes for the problem.
4. **Improve** the process. Generate and evaluate solutions to improve the process. Select solutions that result in the greatest benefits.
5. **Control** the new process and frequently take measurements to ensure consistency and to track progress. Do not just monitor, take active control if necessary.

Additional details for each of the DMAIC phases can be found in a number of Six Sigma texts (see for example, Pande, Neuman, & Cavenaugh, 2002), along with suggested tools to accomplish each step. In practice, there is some overlap across phases because of the iterative nature of the Six Sigma methodology. Not all tools are necessarily used in every project, but the standard Six Sigma toolkit is large and versatile and is available to every Six Sigma practitioner.

Organization and Staffing

In order to take full advantage of Six Sigma methodology, company-wide support is required, as well as extensive training and involvement at many levels. The following is a description of the roles necessary for successful implementation of Six Sigma across a company:

1. **The Executive Leader** is the driving force behind adoption of the Six Sigma philosophy for the whole company. Typically the CEO or president, this leader visibly demonstrates personal commitment to the program, provides inspiration and motivation to the other participants, and engages other members of senior management on the Six Sigma journey.

2. **The Master Champion or Executive Sponsor** must be highly visible and chosen by the Executive Leader. This person oversees the company-wide implementation of Six Sigma and facilitates any cross-functional projects.

3. **Deployment Champions** select projects from their divisions, provide leadership and commitment, and work to implement Six Sigma throughout their divisions or subsidiaries. These are often division vice presidents.

4. **Project Champions** participate in project selection and oversee, support, and fund specific projects and personnel. They choose, evaluate, and support Black Belts and Green Belts, identify and remove barriers, help teams attain their resource commitments, and have the necessary authority to make the recommended process changes happen.

5. **Master Black Belts** are highly experienced Six Sigma leaders. They guide, train, and support certified Black Belts and Green Belts and staff being trained in Six Sigma methodology and provide the links between the Six Sigma organization and management. Master Black Belts work full time supporting several Six Sigma teams.

6. **Black Belts** have been trained and certified in statistical Six Sigma tools following successful completion of several Six Sigma projects. They oversee projects being conducted by less experienced certified staff and their teams and provide advanced methodological support when needed. Black Belts can also work full time supporting projects.

7. **Green Belts** lead teams to work on specific projects to solve real problems. Green Belts undergo basic DMAIC methodology training and after successful completion of a project, are staff certified as Green Belts. Green Belts are leaders in the business and typically spend a day or two a week on Six Sigma projects. To advance to the level of Black Belt requires additional training and completion of more complex projects.

8. **Cross-Functional Project Teams** are composed of staff with extensive knowledge of the process under study. In some companies, these team members are identified as Yellow Belts to emphasize the critical role they have in process improvement.

There is a wealth of more detailed information available on the Web and in the literature about the Six Sigma approach, methodology, training, and implementation requirements, with examples from a variety of companies (see, for example, www.iSixSigma.com; www.apqc.org;

www.bptrends.com; Chowdhury, 2001; Leavitt, 2001; and Redinius, 2004).

Six Sigma at ETS

Growing competition in the testing industry and the need to produce increasing numbers of tests required by the No Child Left Behind legislation forced ETS leadership to look more closely at their products and processes. They made a commitment to improving quality in existing programs and also recognized that a systematic quality control/improvement methodology needed to be introduced not only to maintain and improve existing programs but also to ensure that ETS could expand in the NCLB market—that is, undertake a lot of new work—and still maintain the highest quality standards. Other process improvement methodologies had been tried previously, but Six Sigma appeared to provide the discipline we needed. The Juran Institute's Program for Performance Improvement was being used, but with few tangible results and little senior management commitment. Several staff had been trained in Michael Hammer's Business Process Re-engineering methodology, but it was piece-meal and staff felt it lacked the human element. At the same time, other staff members were being trained in Total Quality Management tools.

As these various methodologies were being tried in various segments across the company, ETS welcomed a dynamic new president, who had participated in the Six Sigma program at his previous company and had seen the positive impact it had there. He brought along a senior vice president for our Operations Department, who, although trained in Six Sigma, did have initial reservations about the program—mainly concerns that the projects selected for the program would be *made-up work*, rather than serious projects aimed at improving vital functions of the company. Based on the president's and senior vice president's endorsement, the other senior management (vice presidents and executive directors) embraced the concept and became the first Champions. Among the many consultants working at ETS at that time were two Master Black Belts who agreed to develop and run an in-house Six Sigma training program. The first Green Belt class started in September 2001.

The Champions proposed likely projects and Green Belt candidates from across the company to the Office of the President and the ETS Business Council. The final lists of projects were those that would have the most financial impact and could be completed within a 3 to 6 month period of time. Candidates were selected based on their work history, which showed many of the following characteristics:

- Welcomes change and new ideas
- Receptive to and uses feedback
- Self-motivated (does not need reminders)
- Someone who wants to be accountable for work output
- Analytical in approaching problems
- Able to work well with others
- One who does not give up easily (perseveres)
- Establishes standards and metrics around whatever tasks she or he is involved with
- Willing to commit to extra effort (does what it takes)
- Naturally looks for improvements
- Smart/intelligent
- Able to work independently
- Able to work on multiple tasks without sacrificing any
- Some previous knowledge in group dynamics
- Some facility with math

The selected staff were encouraged to participate in the new program, with the benefits to include intensive training of new skills, monetary bonus ($2,000 for Green Belt completion and $3,000 to $5,000 for Black Belt certification), personal growth, formal certification (with a paper certificate and actual green belt plaque), and recognition by their peers through in-house news articles.

A Master Champion was also hired to coordinate the program, to advocate the need for increased quality through the Six Sigma methodology, and to resolve any issues resulting from cross-department conflicts or resistance. This person reported directly to the President, had direct contacts with Senior Management, and had the right personality to move the program along.

Through 2005, six waves of Green Belt groups had undergone training and certification, resulting in over 80 certified Green Belts. Each Green Belt training group ranged from seven to thirteen people. Over the course of three to four months, these groups received three and a half weeks of intensive classroom instruction on the Six Sigma methodology and the DMAIC process, with an additional week of training in the use of MiniTab®, a statistical software package with a rich set of Six Sigma tools (see www.minitab.com).

Each instructional week was followed by several weeks in which each Green Belt trainee worked with his or her project team to proceed in detail through the DMAIC phases. The Master Black Belts provided each Green Belt trainee with a half-day of intensive support to ensure successful progress of the project. On the first day of each instructional week, each candidate made a formal presentation, including sections on the problem, the analysis, the results, the control plan, next steps, and lessons learned, all of which were critiqued by the class, as were their presentation skills. Misdirected or unfocused program statements were refined as a result of these critiques.

Three groups of Black Belt candidates went through two additional weeks of advanced Six Sigma training and completion of more extensive projects, resulting in over a dozen certified Black Belts at ETS. Black Belt candidates were chosen by the Master Black Belts from the group of certified Green Belts, based on the strength of their Green Belt projects, a match between their skills and the requirements of proposed projects, and in consultation with the appropriate Champions.

Because the support and involvement of Champions is vital to the success of the projects, at the beginning of each Green Belt wave new Champions received 12 hours of training on the Six Sigma approach, proper project selection, and the formulation of concise problem statements. Champions were kept involved by periodically meeting with their Green Belts to discuss project status and to refine the problem statement as necessary. In addition, they participated in the formal presentations by the candidates.

The early groups of candidates focused on concerns in the Operations Department at ETS—call center, answer sheet processing, and test book distribution. Subsequent groups branched out to other areas, including test development, statistical analysis, finance, legal, purchasing, and facilities.

Description of Several ETS Six Sigma Projects

Among the many Six Sigma projects completed at ETS, several can serve as examples to provide a flavor for the types of problems chosen, methodologies used, and results realized.

Inaccurate Forecasting of Testing Volumes

Timely and accurate knowledge of the number of examinees for a given test administration is an essential requirement for decisions made by many departments at any testing company (among them, bulletin and test book printing, answer sheet processing, essay reading, test center administration, and score reporting). One large testing program at ETS found that their volume forecasts for the past several years fluctuated over and under actual administration volumes by more than 10% (for one test, it was close to 30%). In one year, this resulted in an absolute variance of about $4.5 million from the budget for this testing program.

The project team first looked at the existing forecasting process and quickly identified two major problems: last minute walk-in examinees were not taken into account and, more importantly, frequent forecast notices were distributed with volumes based on previous administrations adjusted by *gut-feeling* increases or decreases. In fact, these notices were subsequently forwarded *down the line*, where others would add their own personal adjustments. Because the forecasts were taken as fairly accurate, any slight fluctuation caused a flurry of activity across the affected departments.

Given that registration and actual volumes were available for several past years, the project team decided that a multiple regression approach would be the most applicable. A data matrix was constructed, consisting of registration counts for each of the 63 days prior to each test administration for the past 8 years. In addition, the number of actual test takers for those administrations was added to the matrix.

Use of this multiple regression approach has reduced the forecast variance down to about 5%, 8 weeks prior to test administration. The forecast based on registration counts is considered tactical in nature and provides the highest accuracy rate with less fluctuation between administrations. This more accurate 8-week forecast allows 10 weeks for final space and staff planning by our essay reading office.

Although actual dollar savings are not available, we are now managing our test administrations using real data rather than relying on experience-based adjustments. Wasted materials and effort have been reduced, and we no longer have big surprises in administration volumes.

Bad Payments for Services Rendered

One project identified bad payments for services rendered by our customer service center as a major problem for ETS. Test takers can contact ETS for a variety of services, some of which require immediate payment, such as test registration or requests for additional test score reports. Initial fact-finding revealed that over 89% of bad payments were from credit card declines, compared to those from bad checks or credit card charge backs. The Six Sigma project team then narrowed their focus to the reduction of these credit card declines. Further analysis disclosed that only 5% of our credit card processing was real-time, the remainder being batch processed overnight. They identified the root cause of the problem as the fact that ETS would provide the service before the payment went through because the cards were not verified in real time—and then the card would be denied, so we would not get paid for a service rendered.

More detailed analysis of the data caused the team to direct their efforts to domestic sales for one particular CBT testing program, as this program accounted for the vast majority of problematic credit card transactions. They developed an action plan and secured corporate support, including endorsement by program management and allocation of appropriate staff resources. Their next steps included implementation of real-time credit card processing to other CBT testing programs—domestic and international—and improved tracking and reporting of bad checks.

This project resulted in over $1 million in improved cash flow/recovered revenue; once expanded to other testing programs, annual savings of over $2.4 million are expected.

Overproduction of Test Bulletins

Staff from two testing programs noticed that their ratios of printed test registration bulletins to projected test registrants were 8:1 and 5:1, respectively. The negative consequences of such overproduction included reduced product margins and increased costs for printing, shipping, storage and disposal, and inventory management. Data further showed that upward of 34% of unshipped bulletins had to be destroyed in-house. That was in addition to the unused bulletins discarded at receiving schools.

Through the use of process mapping and detailed data analysis, the team determined that several factors contributed to the overproduction—the growing Web registration trend was not being considered, and the process of advanced canvassing of school volumes had been discontinued. There were also regionally misdirected shipments, unnecessary reprints, and initial distribution overshipments. The proposed multifaceted solution was phased in over several printing cycles. It included steps to better utilize regional volume projections, aggressively promote Web downloads of registration materials, develop a communications plan for bulk recipients, investigate methods to extend bulletin shelf life, and implement Just-in-Time inventory mechanisms.

The savings for one testing program amounted to $700,000 in print costs alone, based on reducing bulletin quantities by 75%. This did not include additional savings gained from reduced costs for shipping, storage, order-fulfillment, and bulletin destruction. The team implemented a monitoring system to track the print-to-registrant ratio, as well as secondary metrics to check for possible unintended consequences, such as increased number of calls to the call center, increased number of shipments due to late orders, and increased number of print reruns.

Rejected Honoraria Payment Vouchers

ETS hires people around the world to administer tests and supervise associates at test centers. During a recent year, 76% of their vouchers requesting payment for services rendered were rejected by our financial system. The Six Sigma project team analyzed the data to focus on the exact type and number of errors made on the paper and Web-based voucher forms. Sixty-eight different defects were identified on the vouchers, with two of those causing 64% of the rejects. Poor voucher design and limitations of the financial software were identified as the major root causes of the problem. Redesign of these has resulted in a savings of more than $60,000, with more to be realized through future changes.

Excessive Use of Warehouse Storage

In 2001, warehouse storage at ETS exceeded 99% of capacity. Using Six Sigma tools, the project team identified several root causes, one of which was the lack of a policy for automatic destruction of materials after an extended period of inactivity. This lack of policy resulted in the average stock age with no activity being close to two years and storage costs for maintaining these items exceeding $619,000 annually. Implementation of a new policy reduced warehouse usage to 86% of capacity, eliminated the need to utilize an off-site storage site for the first time in 20 years, and reduced off-site storage and transportation costs by $185,000.

Answer Sheet Rejections

One large testing program found that 5% of their returned answer sheets were being rejected, requiring expensive manual resolution. Through extensive data analysis, the project team found that 33% of the rejections were caused by incorrect input of test taker names. Additional rejections resulted from the answer sheets being processed before the test registration forms. By changing the instruction on the answer sheet for entering the name and by changing the process flow so registration forms are processed before answer sheets, the project produced a $142,000 reduction in labor costs dealing with the rejections.

Conclusions and Lessons Learned

The implementation of Six Sigma methodology at ETS has not been completely problem-free. Although voicing initial support, some Champions invested very little time toward helping Green Belt candidates with their projects or toward resolving interdepartmental conflicts over proper procedures or priorities. Candidates and their team members were often obligated to fit project work into their otherwise full-time commitments. Sometimes when a computer software solution was proposed, existing IT commitments or financial and staffing barriers prevented timely implementation of the software. In addition, some projects were terminated because of the transfer of the Green Belt candidate to another department or to a more important project, difficulty in the collection of accurate and complete data, unfocused problem statements, and decreased motivation of the Green Belt candidate over time.

A new group of candidates, consisting of a dozen vice presidents and executive directors, received Black Belt training in the summer and fall of 2006. In hindsight, senior management should have been our first wave in 2001, in order to have them fully trained in Six Sigma methodology and providing top-down oversight and commitment to all subsequent groups. They also could have provided a continuing pipeline of projects on which new candidates and certified Green Belts could have worked. As it was, most Green Belt candidates completed only their one project required for certification and no more. Personal momentum was lost, as were opportunities to spread Six Sigma methodology into other areas of ETS. In a related vein, numerous projects were not fully completed at the time of the final formal presentation in class, and there was no follow-up provided to support their eventual completion.

In the overall balance, however, it is fair to say that in the relatively brief period of time since ETS adopted this program, we has realized documented savings of $10 million and expect to save considerably more as additional problems are defined and eliminated with this valuable set of tools. Additionally, more ETS staff are aware of the benefits of cross-functional teams, more decisions are based on data rather than opinion, and the Six Sigma approach is spreading across the company.

References

Biolos, J. (2002). Six Sigma meets the service economy. *Harvard Management Update*, No. U0211A.

Chowdhury, S. (2001). *The power of Six Sigma: An inspiring tale of how Six Sigma is transforming the way we work*. Chicago: Dearborn Trade.

General Electric. (n.d.). *What is Six Sigma? The Roadmap to Customer Impact*. Retrieved October, 5, 2005, from http://www.ge.com/sixsigma

Leavitt, P. (Ed.). (2001). *Deploying Six Sigma to bolster business processes and the bottom line*. Houston, TX: American Productivity and Quality Center (APQC), www.apqc.org

Pande, P. S., Neuman, R. P., & Cavenaugh, R. R. (2002). *The Six Sigma way: Team fieldbook*. New York: McGraw Hill.

Pyzdek, T. (December, 1997). *Motorola's Six Sigma program*. Retrieved March, 19, 2002, from www.qualitydigest.com/dec97/html/motsix.html

Redinius, D. L. (December, 2004). The convergence of Six Sigma and process management. *BPTrends*. Retrieved February 1, 2006, from www.bptrends.com

Six Sigma DMAIC Process. (n.d.). Retrieved April, 30, 2006, from http://www.isixsigma.com/library/content/c020617a.asp

15
Using Data Forensic Methods to Detect Cheating

David Foster
Kryterion

Dennis Maynes
Caveon Test Security

Bob Hunt
Cisco

Introduction

Why do employers encourage skill certification? Why do schools and colleges in many countries use tests for educational admission and advancement? It is because they value objective measurements of human knowledge, skills, and abilities. The rapidly expanding use of tests in industry and commerce further indicate that testing is not only valued, but trusted.

That trust is the result of experience with tests that provide consistently useful results. To preserve that stakeholder trust, painstakingly won over a long period of time and effort, testing programs must be attentive to old and new threats to a key, and often neglected aspect of testing: security.

The purpose of this chapter is to explain how test-response data analysis (*data forensics*) can be used to identify test cheating and piracy, and to direct remedial action. The chapter describes the size and forms of the cheating problem, and provides illustrations using real test data of how data forensics can be confidently used to detect and reform cheating.

The Cheating Problem

In just the brief span of time since the formation of Caveon Test Security (2003), enough experience has been gained to have formed and confirmed opinions about the scale and effectiveness of cheating and test piracy. Those opinions can be summarized as follows: *cheating and test piracy are prevalent, and usually successful.*

These observations are easily confirmed by other trends including the growing numbers of students who admit to cheating in year-to-year surveys (McCabe & Bowers, 1994; Josephson & Mertz, 2004). In summarizing much of this research, Cizek (1999) wrote that that the trend studies all agreed that the proportion (of cheaters) is high and not going down. Evidence of the growth of cheating can also be found on the Internet where protected test materials are sold or shared (Foster & Zervos, 2006).

These and other sources suggest that the cheating problem has grown with the popularity of tests to certify knowledge, skills, and abilities in almost every area of endeavor. Cohen and Wollack (in press) state that with more and more tests serving as the gatekeepers to many professional goals, the motivation to cheat continues to increase.

Much has also been said about a broader erosion of ethics which is otherwise illustrated by drug scandals in professional sports, corporate fraud, "aggressive" tax strategies, resume fraud, and so forth (Callahan, 2004). Increasingly, "situational" factors such as "retaliatory cheating" increase the prevalence of cheating by providing examinees with easy and convenient justifications. Poor quality tests, for example, with too few or too many questions, arbitrary pass/fail thresholds, irrelevant items, and so forth, can trigger situational cheating.

Finally, it must be acknowledged that the Internet is an especially potent development for the distribution and ultimate use of stolen test information. *Braindump* is Internet vernacular for Web sites that harvest test material from examinees and others, and actively market it to aspiring examinees. The material available on many of these sites is often an exact replica of protected tests and items, and can show up within just a few days or weeks after the publication or administration of a test (Naglieri et al., 2004). Internet discussion forums, bulletin boards, and chat rooms provide other connections where test content is often casually shared and discussed. Unfortunately, the disclosure of only a few test items is enough to affect the fairness and validity of a test, and the reliability of test scores.

What exactly is cheating? Cizek (1999) defines cheating as an attempt, by deceptive or fraudulent means, to represent oneself as possessing knowledge. Practically speaking, that means anything that creates or confers an unfair advantage on a test. This is a broad topic that can be discussed under three large subheadings: copying, collusion, and the use of prohibited materials.

- *Copying* by replicating another examinee's answers endures as a preferred method of cheating on paper-and-pencil tests, but its effect has been substantially diminished by the use of computers to deliver test items randomly or uniquely for each examinee. However, large-scale copying is possible where examinees can collect and quickly disseminate test information for later use on the same test. On-demand computerized tests may be especially susceptible, as Potenza (2002) warns, because such testing is vulnerable to test theft. To address this threat, and to preserve the ease and convenience of computerized testing, Davey and Nering (2002) propose that some of the learning over the years in making conventional high-stakes tests secure must be updated to address the new security problems of computerized administration. In the absence of an immediate solution, some large-scale testing programs have delayed using computers to test or have reverted to manual test administration (i.e., paper-and-pencil).
- *Collusion* differs from copying by the involvement of more than one person in an effort to *beat the test* either by impersonation (i.e., proxy testing), by trafficking protected test information (e.g., the sale and sharing of test information on the Internet), or by communication during a test (e.g., coaching and wireless text messaging).
- As a general matter, the *use of prohibited materials* by examinees involves prior access to protected test material. In addition to memorization, knowledge of that material is used on cheat sheets and more recently, in calculators, cell phones, and other devices programmed with reference information.

Policing high-stakes tests for these types of cheating, as well as efforts to steal or pirate test information for personal and financial gain, is a huge and typically under-funded proposition. This is because the current methods of policing cheating are expensive and incomplete. Even with video monitoring, proctors can only intermittently observe one or a few of the most obvious cheaters. With such limited ability to detect cheating, most testing programs remain blissfully unaware of important security risks, often to the detriment of their reputations and investments.

Data Forensics

As used in this chapter (and increasingly more widely), the phrase *data forensics* is meant to describe an investigative methodology based on the statistical analysis of test and item responses for indications of cheating and piracy. These test and item responses provide several types of observational evidence including item response patterns, answer-change patterns (*erasures*), item response times, as well as test retake and examinee ability test patterns.

The purpose of data forensics is to collate and analyze this observational evidence for the ultimate purpose of making inferences about the presence or absence of cheating within groups of examinees and individual test results. The statistical analyses are carefully selected and calibrated for sensitivity to patterns regarded as indicative of cheating and piracy, with low probabilities of occurrence by chance alone.

Like other detection methodologies using statistics, data forensics is susceptible to type I error (falsely detecting cheating) and type II error (failing to detect cheating). Generally, practitioners must identify an acceptable level of type I error that will make the output of the analysis more or less conservative and yields more reliable results (type II errors are much more difficult to model and calibrate).

The data forensics methodologies described in Table 15.1 differ by type of data analyzed and the use of different statistical models including, for example, copying and erasure models. Each methodology can be used to evaluate individual tests as well as groups of tests for evidence of cheating.

The power of these methodologies to detect anomalies from comparisons of test data with related statistical models increases with sample size. Similarly the strength of inferences of cheating drawn from those anomalies is increased where there is no apparent environmental explanation such as illness, or an interruption or disruption during the testing event.

Types of Cheating Addressed by Data Forensics

Erasures and Answer Changing

The most typical form of what is commonly referred to as erasure analysis consists of counting total answer changes and/or wrong-to-right

TABLE 15.1   Data Forensics Methodologies

Data Forensics Methodology	Data	Security Risk(s) Addressed	Description of Computation
Erasures and answer changing	Item responses	Coaching by proctors or test administrators; posttest answer-sheet tampering	Summary counts of wrong-to-right erasures and other erasures
Collusion	Item responses	Answer-copying; coaching; proxy testing	Answer-copying statistical index such as omega or g2
Gain scores and volatile retakes	Test events and scores	Prior access to protected test content; answer-copying; piracy	A gain score model that predicts expected gains using test scores from previously taken tests or related subject areas
Retake policy adherence	Test events and scores	Violations of program specific retesting policies	Testing program-specific rules that tabulate whether a test was attempted in violation of a specific program policy
Response aberrance	Item responses	Prior access to protected test content; piracy	Variations of person-fit statistics or models that are designed to measure test-taking inconsistency
Latency aberrance	Item latencies	Prior access to protected test content; piracy; proxy testing	An appropriate prediction model (such as linear regression) that accounts for item difficulty and an examinee's mode of working

answer changes on an answer sheet (although this is also possible with computerized tests) and then determining whether the counts are anomalously high in relation to other examinees of similar ability.[1] It is important to observe that examinees of approximately average ability tend to change answers more often than examinees at higher levels of ability.

At Caveon, several models for erasure analysis have been investigated and employed, depending upon the type and availability of test data. For example, if item-level erasure data are available, then individual binomial or multinomial logit models are devised. If only summary counts are available, then aggregate models will be used. Generally, better models result when item-level variability and the categories of erasures can be explicitly modeled.

The simplest model is based upon the binomial distribution and only counts whether an erasure or a wrong-to-right erasure was observed. If the data are provided with erasure categories then minimally a trinomial distribution should be used with categories of wrong-to-right erasures, other erasures, and no erasures. Erasing and answer changing, however, are such low frequency activities that normal approximations are rarely adequate for estimating the required probabilities. Exact binomial and trinomial distributions will generally be preferred for estimating the probabilities of the extremeness of the counts.

When exact trinomial or multinomial distributions are used to model the erasures, the challenge becomes assessing extremeness in the multivariate distribution. As Barnett and Lewis (1994) observed, this situation requires formulating an ordering principle:

> The idea of extremeness still inevitably arises from some form of 'ordering' of the data. What is now lacking is any notion of order, and hence of extremeness, as a formal stimulus to the declaration of an outlier. Although no unique unambiguous form of total ordering is possible for multivariate data, different types of sub-(less than total) ordering principle may be defined and employed. (pp. 269–270)

Barnett and Lewis offer advice on types of ordering that are possible. Generally, the ordering must be consistent with the notion of extremeness that is being employed. Multivariate analysis of the data is very powerful, but it requires additional care to avoid the introduction of inconsistencies in the objective probability principles that data forensics requires.

Collusion

Collusion analysis recognizes that extreme similarity between test records can result from a variety of activities apart from glancing at another examinee's answers, namely, organized cheating. Generally, measures of collusion are premised on the assumption of statistical independence between test responses. When the similarity between two or more test records is extreme, the independence hypothesis is rejected in favor of an alternative hypothesis capable of explaining the relationship.

The best general purpose answer-copying statistics are Wollack's (1997) omega and g2. These statistics however, require identification of a source and copier, and are conditioned upon the observed responses of the source. Because the source and the copier are unknown when this statistic is used as a detection device, two statistics must be computed. The two resulting statistics must then be correlated, but no objective probability principle for this correlation has yet been published.

Instead, we have focused on deriving a collusion detection capability using similarity statistics that do not require the specification of a source and copier. One of these statistics uses an exact trinomial distribution and the other a likelihood ratio test.

Without modification, these statistics require the comparison of every administered test with every other administered test and can quickly exceed the capacity of most computer systems. Our experience however, indicates that the likelihood of collusion decreases dramatically with increases in time and physical distance between examinees. That is, only pairs of test scores associated closely in time or space need to be evaluated.

An important further aspect of this analysis is postprocessing of the detected pairs of highly similar test records in order to evaluate possible relationships. Some of the relationships that have been discovered include pairs of test takers that work together on one test after another, proxy test takers who take exams serially for one individual after another, and coaching by test center staff.

Figure 15.1 shows a cluster of seven tests with different examinee IDs, but similar test dates, time, and scores. The collusion analysis identified the test records as matching too closely to have occurred by chance.

Gain Scores/Volatile Retakes

The purpose of gain score or volatile retake analysis is to identify extreme changes in an examinee's test scores between consecutive administrations of the same test. Because test results are considered a valid reflection of an examinee's ability, the generally accepted

Examinee	Test	Site	Country	Date	Time	Score	Prob	
283	101	Site 12	U.S.	1/15/2003	13:04:18	0.77	246	
405	101	Site 4B	U.S.	5/24/2003	3:17:44 PM	0.87	143	**Possible proxy**
351	101	Site 4B	U.S.	5/24/2003	4:47:03 PM	0.88	258	**test-taker**
860	101	Site 4B	U.S.	5/24/2003	4:12:16 PM	0.88	5029	
446	101	Site 4B	U.S.	5/24/2003	4:48:52 PM	0.85	88	
440	101	Site 4B	U.S.	5/24/2003	5:16:50 PM	0.83	5029	
123	101	Site 4B	U.S.	5/24/2003	4:09:46 PM	0.88	85	
559	101	Site 4B	U.S.	5/24/2003	4:46:14 PM	0.82	85	
756	101	Site 17	U.S.	1/24/2003	2:34:00 PM	0.85	2134	
659	101	Site 17	U.S.	4/11/2003	9:42:30 AM	0.85	2134	

Figure 15.1 Collusion example.

explanation of large score increase between tests is cheating; specifically, prior access to protected test content.

A regression model is usually used as the statistical model for determining whether a test score change (*gain score*) is unusual. When gain scores are analyzed the regression model takes the form of a pre- and posttesting model. Models that are based upon the standard error of measurement or variance estimates of the test scores are also reasonable models. These models should be formulated using a statistical distribution so that the objective probability principle can be applied.

If a pre- and posttest regression model is used, then the standardized residual will follow an examinee's t-distribution under the null hypothesis that the data are adequately described by the regression model. In practice an a priori threshold may be specified for determining if the observed change is excessive. An appropriate adjustment should be made when large numbers of gain scores are evaluated and the procedures are used to detect cheating. When used to detect individual instances of cheating, the gain score analysis will only be useful in detecting the most extreme changes, especially those centered within a group or cluster.

Retake Analysis

Many certification and licensure programs allow tests to be retaken within certain parameters to confirm or improve scores. Programs that offer these *retake* policies, such as a 15- to 30-day waiting period between test administrations, generally regard a violation of the policy as cheating due to the unfair advantage associated with using the test for practice. Retake analysis amounts to identifying rule violations in test event data, and exploring whether retake violations are associated with other statistical indicators of cheating.

Response Aberrance

Response aberrance is a measure of inconsistent test item responses which, depending on the observed patterns, may be indicative of cheating or piracy. The standard of consistency that is applied is usually a parametric or nonparametric test model. The typical response aberrance statistic is a person-fit statistic (or a lack of fit statistic) that is derived from an Item Response Theory test model, including the nominal response model introduced by Bock (1972), which estimates the likelihood of different multiple-choice answer options based upon an examinee's ability or proficiency.

Cheating detection using person-fit statistics, however, is difficult for at least two reasons: (1) person-fit statistics measure negative attributes (i.e., nonconformance to a test model); and (2) many behaviors in addition to cheating can inflate person-fit values. These problems can be surmounted by introducing the assumption that a cheater's response behavior or *ability* markedly varies when responding to test items about which they have prior knowledge.

The resulting *bimodal* test-taking model (i.e., high ability on some items and low ability on others) is a mixture model which estimates the difference between two ability levels. A likelihood ratio statistic is used to measure strength of the fit and provides the foundation for inferences about whether an examinee's responses are aberrant.

Figure 15.2 shows the item response patterns for five individuals with low overall test scores. Bimodal aberrance analysis of these test records

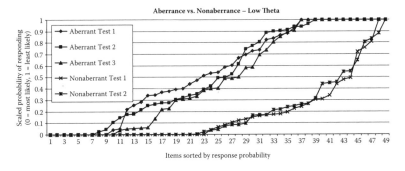

Figure 15.2 Example of response aberrance.

indicates that three of these tests are aberrant due to the improbability of item responses at every ability level. Although achieving the same score, three examinees consistently select highly unlikely answers (in relation to their demonstrated overall ability). Because many item responses indicate lower ability than the overall demonstrated level of ability, the chief concern in this case is theft of test items or *piracy*.

Latency Aberrance

When the amount of time taken by an examinee to answer a test question (item response time) is recorded (using for example, a computer-administered test), the resulting values can also be analyzed for inconsistent response patterns indicative of cheating. In our experience, latency aberrance analysis is uniquely powerful for detecting piracy.

Response latency models are not well established; there is little prior research on how examinees use time in responding to test items, although models do exist which describe expected latencies in relation to certain item attributes such as item complexity and difficulty. These existing models can be used to empirically derive the behavioral models needed to analyze latency aberrance, including working speed and ability.

Figure 15.3 shows a high-score test record with response times too brief to allow the examinee to respond to the items in a normal way. On several items that should have taken over 2 minutes to complete (based on a comparison with average latencies for all examinees), the examinee

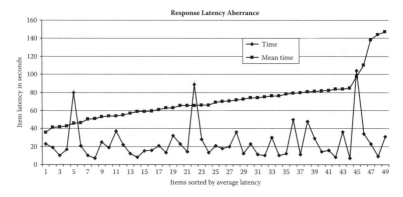

Figure 15.3 Example of latency aberrance.

took only a few seconds. The examinee's latencies were uncorrelated (tau = –.04) with the average latencies for all the test takers.

Combining Evidence

Each data forensics statistic is specifically designed to measure one or another aspect of cheating. If the output of any one statistic is extreme then no other information may be required to support a determination of cheating. Often however, the output may not be extreme, but because the statistics represent independent dimensions of cheating (under the null hypothesis of no cheating) they can be combined to increase detection power.

Generally, if the probability can be evaluated for the probability distribution of the statistic, and if the statistic is derived such that an upper tail test is appropriate, a transformation exists to convert the statistic into a chi-square statistic. Under the assumption of independence the transformed data forensics statistics can be summed and tested using a chi-square distribution.

It is especially essential to combine evidence when analyzing groups of tests arising, for example, from a classroom or testing site. Testing a sample of tests for cheating is identical to testing a sub-sample for the presence of statistical contamination.

The most direct approach counts the number of outliers within the sample and then tests the hypothesis that the number of outliers is consistent with the expected number of outliers for the group. When counts are used, binomial distributions and mixtures of binomial distributions become the modeling distribution of choice. These distributions provide the objective probability principle that is needed to assess extremeness of observed counts.[2]

Effects on Test Performance

Aberrance, collusion, and volatile retake measures can be used to examine the effects of cheating on the performance of the exam, and specifically to answer the question, "is the test performing as designed?"

Figure 15.4 depicts a strong relationship between aberrance and changes in test scores. In this case, an increase in aberrance over a 4-week period is directly related to a real decrease in test scores over that same period.

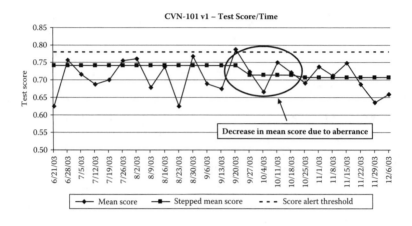

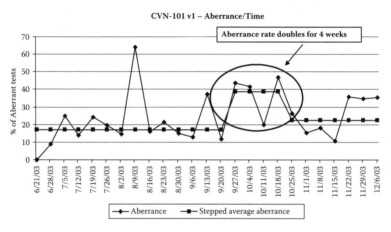

Figure 15.4 Aberrance and score change.

These analyses provide more than a snapshot of score drift at a particular point in time, they show a method of continuously monitoring exam and item performance, a task critical when electronic testing is used.

Practical Considerations

For data forensics to work effectively, testing programs should ensure that the needed test-response data are collected, validated or verified, and easily used. This section describes some of those data requirements in greater detail.

- *Location* includes information about the geographic location of the testing site, staff (administrators and/or proctors) identifiers, and operating hours.
- *Test and Item Information* includes test version and question identifiers as well as examinee item responses. The registration number of the test event and the order in which the items were presented to the examinee (if available) can also be useful.
- *Latency* information for each item as well as any review time used by the examinee (if applicable). Other important temporal test information includes the test date and the times when the test began and ended. (Note: The sum of all item latencies should equal the total time for the test.)
- *Examinee Information* is vital to later remediation activities. This includes name, address, e-mail address, phone numbers, and so forth. A testing ID number, if available, is also very useful.
- *Data Clarity.* Test data should be free from error and contamination. If errors or contamination are present, the data may need to be processed several times before they can be detected and corrected. Information that typically contaminates test data includes: (a) duplicate records of the same test event; (b) failure to remove test records associated with test quality assurance or some other administrative task; and (c) incomplete or confused item response data. The omission of test records also poses important questions about the useable scope of the data.
- *Amount of Data.* Use as much data as possible spanning the life of a test. Quantity improves the reliability of the results. However, Caveon and others have devised methods of performing these analyses on small samples. Longitudinal data makes it possible to evaluate the effects of suspected cheating on the test and specific items over time.
- *Standards.* Adherence to industry data standards for data collection, storage, and retrieval including Sharable Content Object Reference Model (SCORM) and Question and Test Interoperability Standard by IMS Global Learning Consortium, Inc. (IMS/QTI) make data forensics analysis more straightforward, less expensive and if necessary, allow data to be combined across programs.

Defensibility of Evidence from Data Forensics

The information value of data forensics is only as great as a testing program's ability to legally use the information for its protection. In the United States, fortunately, there are powerful examples of the legal defensibility of data forensic evidence in both state and federal jurisprudence.

To reiterate, data forensics findings are generally expressed as probabilities that an examinee's test response patterns agree with one or more statistical models of cheating, and in some cases, test theft. Where they have been reviewed by courts, these probabilities have been viewed as credible evidence of some form of testing irregularity (including cheating), giving testing programs a reasonable basis to doubt the validity of a given test result.

One of the more recent cases to reach a U.S. Court of Appeals provides a study in the defensible use of gain score, copying and erasure analyses, and the value of enabling legal preparations such as a binding test-use agreement. *Murray v. ETS* (1999) involved a Louisiana high school student whose score change between his first and second administrations of the SAT led ETS to withhold and ultimately cancel the second score.

Murray, who had been promised a basketball scholarship by the University of Texas-El Paso, failed to achieve a combined score of 820 needed to receive the scholarship on his first administration of the SAT. He consequently enrolled in a four-week "testbusters" course and on his second administration of the test, achieved a score of 1300.

ETS's routine gain-score analysis flagged Murray's score for further investigation using copying and erasure analyses. As reported in *Murray vs. ETS* (1999), statistical analysis showed that the number of Murray's incorrect answers matching the incorrect answers of another test taker (identified as B) could be expected to occur only three times in comparing 100 million pairs of answer sheets. ETS also conducted an *erasure analysis*, which showed a substantial number of erasure marks on Murray's answer sheet where answers apparently had been changed to match answers on test taker B's answer sheet.

These findings might have been used to objectively demonstrate that Murray had in fact cheated, but the legal significance of the case is the way in which ETS used the information and the court's response.

Prior to Murray's June 1996 administration of the SAT, ETS required all examinees to review and sign the SAT registration bulletin before taking the test. This test-use agreement between ETS and each examinee indicated that ETS would review irregular scores, and could withhold or cancel any score if ETS believed that there was misconduct, if there was testing irregularity, or if ETS believed that a reason existed to question the score's validity (*Murray v. ETS*, 1999).

Without direct evidence (i.e., eyewitness evidence) that Murray had cheated, ETS canceled Murray's test score on the basis that the data forensic evidence provided "reason to question the score's validity" and invited Murray to retest and present other evidence supporting the validity of the second test result (*Murray v. ETS*, 1999).

When the ensuing dispute finally arrived before the Court of Appeals, the court agreed that the relevant question was not whether the evidence demonstrated that Murray had in fact cheated, but whether the evidence sufficed under the terms of the test-use agreement to support a score cancelation, and whether ETS had otherwise fulfilled its contractual duty to investigate the validity of Murray's scores in good-faith. ETS fulfilled that duty, the court concluded, by providing substantial evidence regarding its reasons for questioning Murray's scores and by allowing Murray to present evidence supporting his scores, informing Murray of his right to seek independent review, and ultimately allowing Murray to retake the test (*Murray v. ETS*, 1999).

For testing programs interested in deploying data forensics, the lessons of Murray are: (a) the importance of a test-use agreement which defines how data forensic evidence can be used; and (b) the sufficiency of data forensic evidence showing only irregularity to support a score cancellation under the terms of appropriate test-use agreements.

Conclusions and Lessons Learned

The use of technology to make tests more secure has lagged behind the willingness of examinees to use technology for the opposite purpose. Variations of many of the analyses described in this chapter have been in limited use by some testing programs for nearly two decades. Most programs, however, continue to rely on security methods conceived for paper-and-pencil testing to combat current security problems such as the rapid spread of protected test information on the Internet and its use by examinees.

As this chapter explained, data forensics can reveal testing behaviors that are invisible to other types of surveillance such as wireless communication between examinees, or prior access to protected test material on the Internet. Without an equally effective detection strategy, many testing programs (especially information technology skill certification programs) have suffered serious erosions of their credibility, and consequently, their value. Left unchecked, these security failures could threaten the future of testing itself.

Tests have for many years operated as the standard for individual claims of knowledge, skill, and ability. As more testing programs realize the scope and gravity of cheating accordingly, they either need tools to reverse the corrosive effects of the cheating culture on tests, or an alternative system of measurement.

Caveon has assisted many types of testing programs to expand their detection capability with data forensics. Uniformly, these analyses have revealed levels of suspected cheating sufficient to prompt additional efforts to preserve the ongoing security and validity of those important testing programs.

With a sympathetic legal climate, the largest remaining challenge is to move data forensics closer in time to the testing event such that cheating can be detected and stopped as it happens. When that happens, data forensics may eclipse every other form of test security and act as a powerful deterrent to cheating.

Notes

1. Often an assumption of normality (or near-normality) is used to justify the four standard deviation rule as a measure of extremeness. These data are not normally distributed and if the erasures are categorized by type then the data are distributed as sums of multinomially distributed random variables (assuming independence of erasures under the null hypothesis of normal test taking).
2. Appropriate adjustments must be made to protect against alpha inflation.

References

Barnett, V., & Lewis, T. (1994). *Outliers in statistical data* (3rd ed.). Chichester, UK: John Wiley & Sons.

Bock, R. D. (1972). Estimating item parameters and latent ability when responses are scored in two or more nominal categories. *Psychometricka, 37,* 29–51.

Callahan, O. (2004). *The cheating culture: Why more Americans are doing wrong to get ahead.* Orlando, FL: Harcourt.

Cizek, G. J. (1999). *Cheating on tests: How to do it, detect it, and prevent it.* Mahwah, NJ: Lawrence Erlbaum Associates.

Cohen, A. S., & Wollack, J. A. (2006). Test administration, scoring and reporting. In R. L. Brennan (Ed.), *Educational Measurement,* (4th ed.). (pp. 355–386). Westport, CT: Praeger.

Davey, T., & Nering, M. (2002). Controlling item exposure and maintaining item security. In C. N. Mills, M. Potenza, J. J. Fremer, & W. C. Ward (Eds.), *Computer-based testing: Building the foundation for future assessments.* Mahwah, NJ: Lawrence Erlbaum Associates.

Foster, D. F., & Zervos, C. (2006). *The big Internet heist.* Poster presented at Association of Test Publishers annual conference.

Josephson, M., & Mertz, M. (2004). *Changing cheaters: Promoting integrity and preventing academic dishonesty.* Los Angeles: Josephson Institute of Ethics.

McCabe, D. L., & Bowers, W. J. (1994). Academic dishonesty among males in college: A thirty year perspective. *Journal of College Student Development, 35*(1), 5–10.

Murray v. Educational Testing Service, 170 F.3d 514, (5th Cir. 1999).

Naglieri, J. A., Drasgow, F., Schmit, M., Handler, L., Prifitera, A., Margolis, A., & Velasquez, R. (2004). Psychological testing on the Internet: New problems, old issues. *American Psychologist, 59*(3), 150–162.

Potenza, M. (2002). Test administration. In C. N. Mills, M. Potenza, J. J. Fremer, & W. C. Ward (Eds.), *Computer-based testing: Building the foundation for future assessments.* Mahwah, NJ: Lawrence Erlbaum Associates.

Wollack, J. A. (1997). A nominal response model approach for detecting answer copying. *Applied Psychological Measurement, 21*(4), 307–320.

16

Detecting Exposed Test Items in Computer-Based Testing*

Ning Han and Ronald Hambleton
University of Massachusetts at Amherst

Introduction

Exposed test items are a major threat to the validity of computer-based testing. Historically, paper-and-pencil tests have maintained test security by (a) closely monitoring test forms (including their development, printing, distribution, administration, and collection), and (b) regularly introducing new test forms. Because, however, of the necessity of daily exposure of an item bank to examinees in a computer-based testing environment, standard methods for maintaining test security with paper-and-pencil administrations are no longer applicable. Failure to adequately solve the item security problem with computer-based testing almost certainly guarantees the demise of this approach to assessment. Detecting particular examinees who may have benefited from exposed test items is another direction being developed currently by researchers, but this direction is not considered further in this chapter (see Karabatsos, 2003).

Most of the research to date for limiting item exposure with computer-based tests has focused on finding ways to minimize item usage: Expanding the number of test items in a bank (either by hiring extra item writers and/or using item generation forms and algorithms; see Impara & Foster, 2006; Pitoniak, 2002), establishing conditional item exposure controls (see for example, Revuelta & Ponsoda, 1998; Stocking & Lewis, 1998; Yi & Chang, 2003), rotating item banks

* Center for Educational Assessment Research Report No. 526. Amherst, MA: University of Massachusetts, Center for Educational Assessment.

(i.e., substituting one item bank for another), expanding the initiatives to reduce sharing of test items on the Internet (see, for example, Impara & Foster, 2006), shortening test administration windows (with the intent of reducing the time period when exposed test items can be shared among examinees), modifying the test design (with the intent of reducing the number of items that examinees are administered, without loss of precision (see for, example, the work of Jodoin, Zenisky, and Hambleton, 2006), implementing an item bank inventory (see Ariel, van der Linden, & Veldkamp, 2006), and better item bank usage (Yi & Chang, 2003).

A different approach for addressing the problem is to focus attention on the generation and investigation of item statistics that can reveal whether test items have become known to examinees prior to seeing the items in the test they were administered (Han, 2006; Han & Hambleton, in press; Lu & Hambleton, 2004; Segall, 2001; Zhu, Yu, & Liu, 2002). If any exposed items can be spotted statistically, they can be deleted from an item bank. Along these lines, several item statistics have been proposed (see, for example, Han, 2003, 2006; Lu & Hambleton, 2004).

Han (2003, 2006) proposed the concept of *moving averages* for detecting exposed test items. The moving average is a form of average which has been adjusted to allow for periodic and random components of time series data. A moving average is a smoothing technique used to make the long-term trends of a time series clearer. Much like moving averages that are used on Wall Street to monitor stock price changes and in manufacturing industry to control product qualities, item performance can be monitored over time (e.g., after each item administration), and any changes can be noted and used to identify potentially exposed test items. Preliminary research was encouraging (see Han, 2003). At the same time this research was based on the assumption that the examinees' ability distribution over time was stationary and a simple item exposure model was put in place. Several directions seemed worthy of follow up research: investigating additional item exposure statistics, and evaluating these statistics under different conditions such as with shifting ability distributions over time and with various types of items (e.g., hard and easy, low and high discrimination), and for several exposure models.

More specifically then, the purposes of the research described in this chapter were (a) to evaluate several item exposure detection statistics in the presence of shifts in the ability distribution over time, (b) to address the suitability of the item exposure detection statistics under a number of item exposure models, and (c) to investigate item exposure detection for items with different statistical characteristics. The first purpose was essential because it simply is not reasonable to assume a fixed ability distribution at all times during a testing window. Some drift in the distribution might be expected—for example, the poorer examinees may come first, and higher ability examinees may follow later in the window. Several new item exposure statistics needed to be investigated because the moving p-value statistic that Han (2003) considered was sensitive to ability shifts and therefore, it was less suitable for use by testing agencies doing computer-based testing: Shifts in ability distribution and detection of exposed items using moving p-value averages are confounded. Although it may be true that the ability distribution of examinees will by-and-large be equivalent over time with many tests, we felt that item exposure detection statistics that are free of this questionable assumption deserved study.

Achieving the second purpose would provide data on competing item exposure detection statistics under various item exposure models. For example, in one simple model, after an item is exposed by an examinee one might conjecture that all examinees have knowledge of the item and answer it correctly if it is selected for administration again. Several other item exposure models needed to be investigated that were more realistic.

The third purpose was added to the study because we expected that the item exposure detection rate would depend not only on the choice of item exposure detection statistic, sample size, and nature of the exposure, but would also depend on the statistical characteristics of the exposed test items. For example, we expected it would be very difficult to detect exposed items when they were easy for examinees (after all, examinees are already expected to do well, and any improvements in item performance due to exposure then would be small); harder items should be considerably easier to spot because the shifts in item performance due to exposure are likely to be greater.

Research Design

A great number of simulated data sets were considered in the study. Variables under study included (a) ability distribution (fixed or variable),

(b) type of item exposure model, (c) choice of item exposure detection statistic, and (d) statistical characteristics of exposed test items.

The level of item exposure was controlled in the study by one parameter, ρ, and it was varied from no exposure ($\rho = 0$) to full exposure ($\rho = 1$) to either 10% or 100% of the examinees. An intermediate value of $\rho = .25$ applied to either 10% or 100% of the examinees was also considered in the simulations.

The study was implemented as follows:

1. A linear test consisting of 75 items with item parameters consistent with item statistics in a national credentialing exam was simulated. So as to roughly approximate the actual testing condition, we considered an item administration level of about 20% to examinees (i.e., items were administered to about 20% of the examinees). Because the proposed item exposure detection statistics monitor examinee's response on an item over time, it is independent of the delivery mechanism of the test. Therefore, a simple linear test design was used without loss of generality of the findings.

2. The number of examinees used in the study was 5,000. We assumed 25,000 examinees in a testing window, with a 20% administration level, so up to 5,000 examinees would see any set of 75 items. Three different ability distributions for the 5,000 examinees were considered: Normal (0.0, 1.0), drifting from a lesser ability group to a higher ability, $\theta \sim N(-1.0+i/2,500, 1.0)$, with i taking on values from 1 to 5,000, and an abrupt shift from $\theta \sim N(-1.0,1.0)$ for the first 2,500 examinees and $\theta \sim N(1.0,1.0)$ for the next 2,500 examinees. In simulating drift, we were assuming that the poorer examinees, generally, would take the test early (average ability = -1.0) and then gradually the ability distribution would shift from a mean of -1.0 to a mean of 1.0 by the end of the testing window. With the abrupt shift in ability distribution condition, after the first 2,500 examinee abilities were sampled from a N($-1.0,1.0$), for the last 2,500 examinees, examinee abilities were sampled from a N(1.0, 1.0) distribution.

3. The probability that an examinee answers an item correctly is

$$P' = P + \rho(1 - P)$$

where

P is the probability of a correct item response computed from the three-parameter logistic IRT mode based on an examinee's ability level and the item statistics ρ is a positive number, $0 \le \rho \le 1$, and was varied in the simulations, to reflect the item exposure model in place

4. Simulation variables in the study were as follows:
   a. Ability distributions
      i. normal
      ii. drifting
      iii. abrupt shift
   b. Extent to which an item was exposed

      $\rho = 0, 0.25, 1$

      $\rho = 0$ was a base-line situation where the item is secure.

      $\rho = 1$ was an extreme situation in which every examinee answers the item correctly.

      $\rho = 0.25$ was a situation where examinee performance, relative to ability and item statistics, was increased to reflect the fact that some general information was being disseminated about the item which gave examinees a boost in their likelihood of success, but not a guarantee they would answer the item correctly.
   c. Statistics
      i. Moving p-values
      ii. Moving averages of item residuals (actual item score—expected item score based on the three-parameter logistic test model)
      iii. Moving averages of standardized item residuals (actual item score—expected item score based on the three-parameter logistic test model/standard error)

         The idea with the item statistics described in ii and iii was to look at item performance compared to expected performance given an examinee's ability estimate. Examinee ability estimates can be calculated after the test administration, and then used along with the statistics for an item and the examinee's item performance to calculate an item residual and the item standardized residual (Hambleton, Swaminathan, & Rogers, 1991). It is only when these differences consistently exceeded what might be expected by chance for the item that the alarm would go off—that is, item exposure would be suspected.
   d. The statistical characteristics of the items

      $b = -1.0, 0.0, 1.0, 2.0$

      $a = 0.4, 0.7, 1.2$

These statistics were crossed to produce 12 item types to focus on in the research—with items ranging from relatively easy ($b = -1.0$) to quite hard ($b = 2.0$), and with low ($a = 0.4$), medium ($a = 0.7$), and high (a = 1.2) levels of discriminating power. These items were embedded in a 75-item test. Item exposure, when it was simulated, always began with the 2,501st examinee in the sequence.

e. Number of simulations for each combination of the previously mentioned situations: 100

f. Detecting exposed test items
Under the no-exposure condition, it was possible for each of the 12 item types to determine the empirical sampling distribution of each of the item statistics after each item administration (100 replications were carried out and the approximate 2.5 and 97.5 percentiles were determined along with the mean of the 100-item statistics). The mean plus two standard deviations and the mean minus two standard deviations were used to approximate the percentile scores. Figure 16.1 shows these values over many item administrations. These extremes were used in the flagging (i.e., detecting of exposed items). Whenever an item statistic exceeded these boundaries, either a type I error was made (if no exposure had been modeled) or exposure was detected (if exposure had been modeled).

Figure 16.1 shows the moving item residual over the number of item administrations, and here, no exposure has been introduced. The horizontal axis is the number of examinees who were administered the test item. The vertical axis is the value of the item residual.

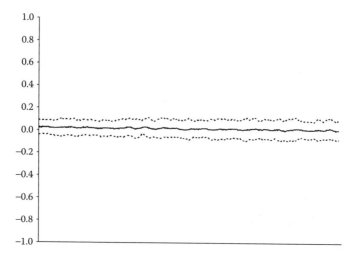

Figure 16.1 Plot of moving item residuals for a nonexposed item showing the average and the 2.5 and 97.5 percentile values obtained from 100 replications.

What is seen is that the plot of the item residual statistic (based on the last 200 examinees) is quite flat and not coming close to the two confidence bands that might suggest the pattern of results is being influenced by unintended exposure of the test item.

A more formal explanation of what is happening follows. Given a sequence of, say, 5,000 examinees:

$$\{\theta_1, \theta_2, \ldots, \theta_t, \ldots, \theta_{5000}\}$$

where θ_t is the true ability of examinee t.

For item i, the binary score for examinee t, t = 1, 5,000 is obtained:

$$\{x_{i1}, x_{i2}, \ldots, x_{i5000}\}$$

Three item statistics (introduced in this study) can be computed and plotted: moving p values, moving item residuals and moving standardized item residuals for the last k examinees (k is referred to as the window size). If k is chosen to be a very low number, such as 25, potential for detecting exposure early is higher, but generally the statistic being plotted would be unstable because it is based upon only 25 examinees. On the other hand, if k is chosen to be very high, say, 500, the item statistic being plotted is stable, but then the item statistic is relatively slow to detect any item exposure. In our research we have found that $k = 100$ is a good balance between instability in the item statistic, and sensitivity to detecting exposure. For example: when the windows size k equals to 100, the sequence of moving p values is:

$$\{p_{100}, p_{101, \ldots,} p_{5000}\}$$

where

$$p_{100} = \tfrac{1}{100}(x_{i1} + \cdots + x_{i100})$$

$$p_{101} = \tfrac{1}{100}(x_{i2} + \cdots + x_{i101})$$

$$\vdots$$

$$p_{n-k+1} = \tfrac{1}{k}(x_{i,n-k+1} + x_{i,n-k+2} + \cdots + x_{i,n})$$

The sequence of moving item residuals is:

$$\{r_{100}, r_{101, \ldots,} r_{5000}\}$$

where

$$r_{100} = \tfrac{1}{100}([x_{i1} - prob(a_i,b_i,c_i,\theta_1)] + \cdots + [x_{i100} - prob(a_i,b_i,c_i,\theta_{100})]$$

$$r_{101} = \tfrac{1}{100}([x_{i2} - prob(a_i,b_i,c_i,\theta_2)] + \cdots + [x_{i101} - prob(a_i,b_i,c_i,\theta_{101})]$$

$$\vdots$$

The sequence of moving item standardized residuals is:

$$\{SR_{100}, SR_{101},\ldots,SR_{5000}\}$$

where

$$SR_{100} = \frac{\displaystyle\sum_{j=1}^{100}(x_{ij} - prob(a_i,b_i,c_i,\theta_j))}{\sqrt{\displaystyle\sum_{j=1}^{100} prob(a_i,b_i,c_i,\theta_j)(1 - prob(a_i,b_i,c_i,\theta_j))}}$$

$$SR_{101} = \frac{\displaystyle\sum_{j=2}^{101}(x_{ij} - prob(a_i,b_i,c_i,\theta_j))}{\sqrt{\displaystyle\sum_{j=2}^{101} prob(a_i,b_i,c_i,\theta_j)(1 - prob(a_i,b_i,c_i,\theta_j))}}$$

For each simulation, we can obtain one sequence for each item statistic. The simulation process itself was replicated 100 times. Therefore, for each item statistic we could obtain 100 sequences. Three new sequences for each item statistic were obtained and plotted too: Mean, Mean + 2SD, Mean – 2SD. For example, for moving p values, the means of the simulations are:

$$\left\{ \frac{\displaystyle\sum_{h=1}^{100} p_{h,100}}{100}, \frac{\displaystyle\sum_{h=1}^{100} p_{h,101}}{100},\ldots, \frac{\displaystyle\sum_{h=1}^{100} p_{h,5000}}{100} \right\}$$

where h stands for the hth replication. And, the 2.5 and 97.5 percentiles of the distribution of the item statistic across the 100 replications can be obtained and plotted too (approximately, mean + 2SDs and mean − 2SDs).

This sequence is plotted in the middle of the plot and the dotted lines are Mean + 2*SD and Mean − 2*SD. The vertical axis is the values of the sequence (or item statistic) and the horizontal axis is the order of the sequence.

Results

Our first task was to determine the window size, that is, the amount of examinee data that would be used in calculating the rolling averages of item statistics. At one point in our research design planning, the choice of window size was going to be a variable in the study, but ultimately we determined from many practice simulations that a window size of 100 was large enough to provide stable statistical information, but not so large that items might go for extended periods without being spotted if they had been exposed. We will leave comprehensive study of the window size variable and its interactions with other variables for another study (see Han, 2006). A review of Figures 16.2 to 16.4 shows the type of variability of these item exposure detection statistics associated with a window size of 100 for item 5 (b = 0.0, a = .7). For the moving average p-values the standard deviation looked to be about .05. For the item residuals, the standard deviation looked to be about .05, and for the item standardized residuals, the standard deviation appears to be about 1.0 (recall that the upper and lower bands cover about four standard deviations).

Comparison of Item Exposure Detection Statistics in the Presence of Ability Distribution Shifts

Figures 16.2 through 16.4 highlight the functioning of the three item statistics for a medium difficult item (b = 0.0, a = 0.7) with three ability distributions—normal, shifting, and abrupt change, respectively. What is very clear is that with a fixed normal ability distribution, all three item exposure detection statistics are quite stable, as they should be—both the item statistics and the 95%

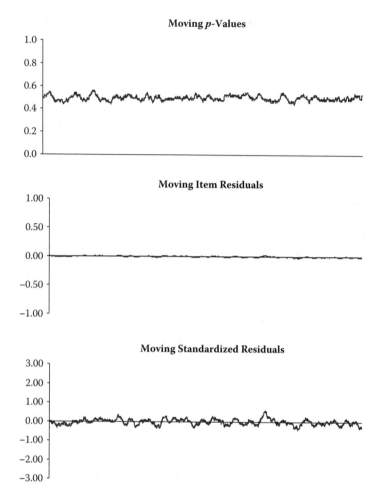

Figure 16.2 Plot of item exposure statistics for item 5 (b = 0.0, a = 0.7). (normal ability distribution, ρ = 0.0, *k* = 100)

confidence bands. With a shift in the ability distribution—gradual or abrupt, the *p*-value statistic shifted too—substantially. Clearly, *p*-value shifts are confounded with shifts in ability distributions and are not reflecting item exposure because there was no exposure. Obviously this finding is not surprising, but the figures do highlight this fact, as well as the stability of the two IRT-based item exposure statistics that take into account examinee ability in their calculations.

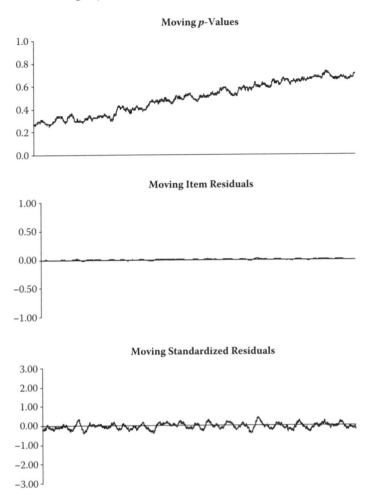

Figure 16.3 Plot of item exposure statistics for item 5 (b = 0.0, a = 0.7). (drifting normal ability distribution, ρ = 0.0, k = 100)

Speed of Detection, Type I Errors and Power of Detection for Items with Various Statistical Properties under Four Exposure Models

Tables 16.1 through 16.8 provide the data we obtained with a constant normal distribution of ability for the examinees over the period of testing. Here, all three item exposure detection statistics were expected to be potentially useful and they were. Table 16.1 shows that with ρ = 1.0, with 100% of the examinees benefiting from the exposed information

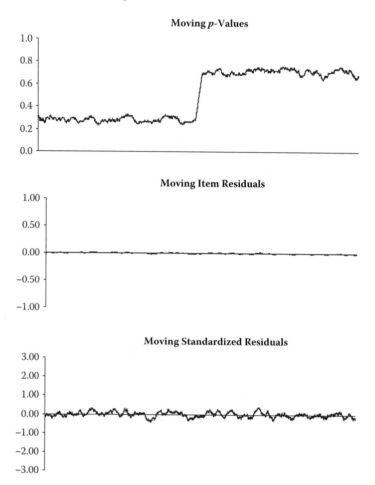

Figure 16.4 Plot of item exposure statistics for item 5 (b = 0.0, a = 0.7). (Abrupt shift in the mean of the normal ability distribution, $\rho = 0.0$, $k = 100$.)

on the 12 items, detection was very fast. Across 100 replications, for example, Table 16.1 highlights that with b = −1.00 and a = 0.4 (a relatively easy and low discriminating item), the average number of examinees who saw the exposed item was 27.4 before the statistic exceeded the threshold. (Note that in the simulations, exposure always occurred with the 2,501st examinee in the sequence of 5,000 examinees who would see the item.) Detection was even faster with harder items. And, in general, more discriminating items were detected faster too, except

TABLE 16.1   Number of Item Administrations, on Average, after Exposure Had Occurred before Exposure Was Detected ($\rho = 1.0$, for 100%, Normal Distribution of Ability)

Item Statistic	Item Difficulty	a = 0.40	a = 0.70	a = 1.20
Moving p-values	b = −1.00	27.4	22.0	28.6
	b = 0.00	15.5	10.4	9.0
	b = 1.00	11.9	7.3	4.5
	b = 2.00	9.2	4.7	2.6
Moving item residuals	b = −1.00	25.3	22.9	24
	b = 0.00	16.3	12.4	11.2
	b = 1.00	12.5	8.7	7.5
	b = 2.00	10.4	6.4	3.6
Standardized item residuals	b = −1.00	25.2	22.6	23.5
	b = 0.00	16.3	12.4	10.9
	b = 1.00	12.4	8.6	7.5
	b = 2.00	10.4	6.7	4.6

when the items were on the easy side. There were very little, if any, differences among the item exposure detection statistics. They all functioned about the same, and well.

Table 16.2 shows the type I and power statistics for the 12 items. Type I errors were based on data compiled from the 1,500th administration of the item to the 2,500th administration. In this portion of the window, there was no item exposure. It is seen in Table 16.2, that under the conditions simulated, the type I error rate varied from 1.5% to 2.7% with the low discriminating items and was somewhat closer to the 5% level with the more discriminating items (2.6% to 4.4% with a = .7, and 1.9% to 6.6% with a = 1.2) which had been the goal. More important in our work was the level of power of detection. In the case with $\rho = 1.0$ and 100% exposure, detection was very easy and the power of detection was 100% for all items. Figure 16.5 shows the findings, graphically, with a normal distribution of ability. More interesting cases follow.

Table 16.3 presents the first set of interesting results for the case where only 10% of the examinees had exposure to the item. Again, the more difficult items are spotted after considerably less item administrations than that of easier items. For example, with b = −1.0, a = 0.4, 320.7 (on the average) examinees were administered the easy item

TABLE 16.2 Type I Errors (I) and Statistical Power (II) ($\rho = 1.0$, for 100%, Normal Distribution of Ability)

Item Statistic	Item Difficulty	a = 0.40		a = 0.70		a = 1.20	
		I	II	I	II	I	II
Moving p-values	b = −1.00	1.50	100.0	3.36	100.0	2.33	100.0
	b = 0.00	2.68	100.0	4.42	100.0	3.60	100.0
	b = 1.00	2.16	100.0	2.86	100.0	5.55	100.0
	b = 2.00	1.99	100.0	4.08	100.0	6.61	100.0
Moving item residuals	b = −1.00	2.14	100.0	2.78	100.0	1.97	100.0
	b = 0.00	2.55	100.0	3.27	100.0	1.94	100.0
	b = 1.00	2.36	100.0	2.56	100.0	2.85	100.0
	b = 2.00	2.02	100.0	2.63	100.0	3.11	100.0
Standardized item residuals	b = −1.00	2.15	100.0	2.78	100.0	2.16	100.0
	b = 0.00	2.55	100.0	3.10	100.0	1.94	100.0
	b = 1.00	2.45	100.0	2.74	100.0	2.88	100.0
	b = 2.00	2.09	100.0	2.59	100.0	2.99	100.0

prior to exposure being detected with the moving p-value item exposure statistic. With the hardest item (b = + 2.0), and with the same item exposure detection statistic, 98.5 (on the average) examinees were administered the item prior to exposure being detected. With the other item exposure statistics, exposure appeared to be a bit quicker. In general, more discriminating items were detected faster than less discriminating items if they were medium to high difficulty.

Table 16.4 shows, for example, that type I errors were in the 1.5% to 6.6% range across all of the combinations of runs. Choice of item exposure detection statistic was of no major significance in the findings. Perhaps the most noticeable result in Table 16.4 is the low power of detection of exposed easy items (b = −1.0 or b = 0.0); 25.2% detection rate was the highest. Whereas for the more difficult items (b = 1.0 and b = 2.0), power of detecting exposure ran as high as 94.7%. Clearly too, for the more difficult items, detection rates were higher for the more discriminating items. For example, considering the most difficult item (b = 2.0), with the standardized item residual statistic, the power rates for items with discrimination levels of .4, .7, and 1.2, were 49.4%, 74.9%, and 93.5%, respectively.

Table 16.5 presents the first set of results for the case where $\rho = 0.25$ and 100% of the examinees had exposure to the 12 items.

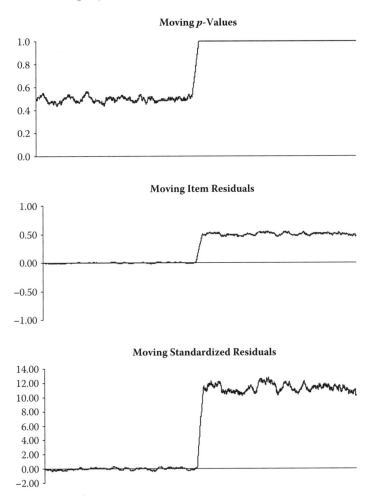

Figure 16.5 Plot of item exposure statistics for item 5 (b = 0.0, a = 0.7). (Normal ability distribution, ρ = 1.0, 100% exposure, k = 100.)

Detection of item exposure did not take very long. Here again, the more difficult items were spotted after considerably fewer administrations than that of easier items. For example, with a = 0.4, 113.5 (on the average) examinees were administered the easy item (b = −1.0) prior to exposure being detected with the moving p-value item exposure detection statistic. With the hardest item (b = 2.0), and with the same item exposure statistic, 39.5 (on the average), examinees were administered the item prior to exposure being detected.

TABLE 16.3   Number of Item Administrations, on Average, after Exposure Had Occurred before Exposure Was Detected ($\rho = 1.0$, for 10%, Normal Distribution of Ability)

Item Statistic	Item Difficulty	a = 0.40	a = 0.70	a = 1.20
Moving p-values	b = −1.00	320.7	301.2	292.3
	b = 0.00	173.8	160.2	140.3
	b = 1.00	169.0	115.9	61.5
	b = 2.00	98.5	57.2	44.8
Moving item residuals	b = −1.00	283.7	313.9	329.8
	b = 0.00	191.7	143.5	188.5
	b = 1.00	140.8	113.8	66.6
	b = 2.00	98.1	61.2	48.8
Standardized item residuals	b = −1.00	283.8	315.4	307.2
	b = 0.00	192.9	149.1	189.5
	b = 1.00	135.0	112.4	67.8
	b = 2.00	96.9	60.8	48.8

TABLE 16.4   Type I Errors (I) and Statistical Power (II) ($\rho = 1.0$, for 10%, Normal Distribution of Ability)

Item Statistic	Item Difficulty	a = 0.40		a = 0.70		a = 1.20	
		I	II	I	II	I	II
Moving p-values	b = −1.00	1.50	8.3	3.36	10.5	2.33	9.8
	b = 0.00	2.68	16.8	4.41	23.7	3.63	25.2
	b = 1.00	2.16	26.6	2.86	38.4	5.55	64.7
	b = 2.00	1.99	47.5	4.08	77.9	6.60	94.7
Moving item residuals	b = −1.00	2.14	9.8	2.78	10.0	1.97	8.7
	b = 0.00	2.55	16.7	3.27	23.8	1.94	24.1
	b = 1.00	2.36	29.5	2.56	41.1	2.84	63.6
	b = 2.00	2.02	49.0	2.62	75.5	3.10	94.0
Standardized item residuals	b = −1.00	2.15	9.9	2.78	10.0	2.16	9.1
	b = 0.00	2.54	16.7	3.10	23.4	1.94	24.3
	b = 1.00	2.45	29.9	2.74	42.2	2.88	63.5
	b = 2.00	2.08	49.4	2.59	74.9	2.99	93.5

TABLE 16.5   Number of Item Administrations, on Average, after Exposure Had Occurred before Exposure Was Detected (ρ = 0.25, for 100%, Normal Distribution of Ability)

Item Statistic	Item Difficulty	a = 0.40	a = 0.70	a = 1.20
Moving p-values	b = −1.00	113.5	123.5	118.8
	b = 0.00	67.9	55.3	55.4
	b = 1.00	53.8	49.7	24.5
	b = 2.00	39.5	21.0	16.1
Moving item residuals	b = −1.00	99.4	119.5	109.9
	b = 0.00	64.9	52.6	56.0
	b = 1.00	47.1	46.2	29.6
	b = 2.00	38.4	23.3	18.7
Standardized item residuals	b = −1.00	99.3	119.1	109.3
	b = 0.00	64.9	52.9	56.0
	b = 1.00	46.3	45.6	29.7
	b = 2.00	38.3	25.1	20.8

With the other item exposure detection statistics, detection of exposure appeared to be a bit quicker, but only marginally. In general, more discriminating items were detected faster than less discriminating items if they were medium to high in their level of difficulty.

Table 16.6 shows, for example, that type I errors were in the 1.5% to 6.6% range as noted before across all of the combinations of conditions. Choice of item exposure detection statistic was of no major significance though the two IRT-based statistics appeared to function a bit better overall. This time, detection rates for exposed easy items ran about 35% to 40%, compared to a detection rate of 100% for the hardest items.

Table 16.7 presents the poorest detection rates of the four item exposure models (ρ = .25, 10% exposure). Even for the most difficult and discriminating items, more than 125 difficult administrations were needed. In the main though, trends were the same: More difficulty and more discriminating items took less time to detect than the easier and/or less discriminating test items.

Table 16.8 shows that the likelihood of detecting exposure was very low. Even for the most difficult and discriminating items, power of detection did not exceed 26%. Choice of item exposure detection statistic was of no major significance.

TABLE 16.6　Type I Errors (I) and Statistical Power (II) ($\rho = 0.25$, for 100%, Normal Distribution of Ability)

Item Statistic	Item Difficulty	$a = 0.40$		$a = 0.70$		$a = 1.20$	
		I	II	I	II	I	II
Moving p-values	$b = -1.00$	1.50	40.9	3.36	39.0	2.33	33.9
	$b = 0.00$	2.68	71.6	4.41	78.0	3.63	85.5
	$b = 1.00$	2.16	88.8	2.86	97.3	5.55	99.8
	$b = 2.00$	1.99	99.3	4.08	100.0	6.60	100.0
Moving item residuals	$b = -1.00$	2.14	46.7	2.78	39.5	1.97	41.0
	$b = 0.00$	2.55	74.0	3.27	80.8	1.94	89.1
	$b = 1.00$	2.36	91.2	2.56	98.2	2.84	99.9
	$b = 2.00$	2.02	99.4	2.62	100.0	3.10	100.0
Standardized item residuals	$b = -1.00$	2.15	47.2	2.78	39.8	2.16	42.0
	$b = 0.00$	2.54	74.0	3.10	80.5	1.94	89.2
	$b = 1.00$	2.45	91.4	2.74	98.3	2.88	99.8
	$b = 2.00$	2.08	99.4	2.59	100.0	2.99	100.0

TABLE 16.7　Number of Item Administrations, on Average, after Exposure Had Occurred before Exposure Was Detected ($\rho = 0.25$, for 10%, Normal Distribution of Ability)

Item Statistic	Item Difficulty	$a = 0.40$	$a = 0.70$	$a = 1.20$
Moving p-values	$b = -1.00$	517.6	473.3	393.2
	$b = 0.00$	530.9	420.6	310.1
	$b = 1.00$	539.2	340.4	186.2
	$b = 2.00$	424.2	173.1	136.8
Moving item residuals	$b = -1.00$	666.5	622.5	721.6
	$b = 0.00$	482.3	538.2	478.2
	$b = 1.00$	558.9	415.1	270.0
	$b = 2.00$	480.9	271.6	179.3
Standardized item residuals	$b = -1.00$	650.5	671.9	674.7
	$b = 0.00$	482.9	591.6	479.0
	$b = 1.00$	573.2	388.0	282.3
	$b = 2.00$	474.4	255.3	180.5

TABLE 16.8   Type I Errors (I) and Statistical Power (II) ($\rho = 0.25$, for 10%, Normal Distribution of Ability)

Item Statistic	Item Difficulty	a = 0.40		a = 0.70		a = 1.20	
		I	II	I	II	I	II
Moving p-values	b = −1.00	1.50	3.34	3.36	4.2	2.33	4.5
	b = 0.00	2.68	4.53	4.41	7.6	3.63	8.0
	b = 1.00	2.16	5.34	2.86	7.7	5.55	16.0
	b = 2.00	1.99	6.81	4.08	15.5	6.60	26.1
Moving item residuals	b = −1.00	2.14	3.72	2.78	3.4	1.97	3.5
	b = 0.00	2.55	4.68	3.27	6.0	1.94	4.8
	b = 1.00	2.36	6.12	2.56	7.4	2.84	10.8
	b = 2.00	2.02	7.03	2.62	11.8	3.10	21.1
Standardized item residuals	b = −1.00	2.15	3.78	2.78	3.4	2.16	3.6
	b = 0.00	2.54	4.67	3.10	5.8	1.94	4.9
	b = 1.00	2.45	6.23	2.74	7.8	2.88	10.6
	b = 2.00	2.08	7.24	2.59	11.5	2.99	20.6

Figures 16.5 through 16.8 highlight the pattern of the item exposure detection statistics for item 5 (b = 0.0, a = 0.7) under the four item exposure models with a normal distribution of ability. What is seen is the following: For $\rho = 1$, and 100% exposure, the item was very easy to detect (see Figure 16.5); for $\rho = 0.25$, 100% exposure, the item took somewhat longer to identify and the power was moderate (see Figure 16.6); for $\rho = 1.0$, 10% exposure, the trend was clear but the item was not identified very often (see Figure 16.7); and finally with $\rho = .25$, and 10% exposure, the exposure was barely detectable in the moving average lines (see Figure 16.8). These figures were presented for illustrative purposes only, and for accurate information on power of detection associated with specific items, see Tables 16.1 through 16.8.

Impact of Shifts in the Ability Distribution

Tables and figures containing the statistical results for the gradually shifting ability distribution and the statistical results for the abrupt shift in ability distributions can be found in Han (2006). All of the findings reported previously for the normal distribution were observed again. The major problem was clear from the levels

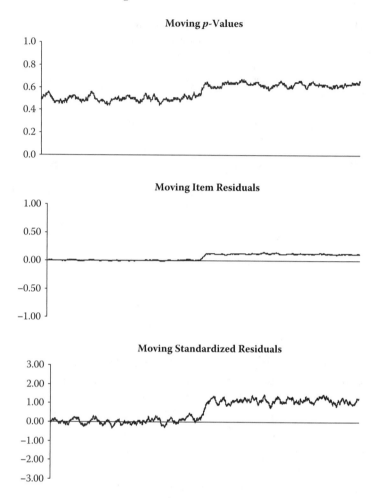

Figure 16.6 Plot of item exposure statistics for item 5 (b = 0.0, a = 0.7). (Normal ability distribution, ρ = 1.0, 10% exposure, k = 100.)

of power of detection with the moving average p-values. These are very high for easy and hard items and both low, moderate, and high discriminating power (and though not reported, but can be seen in Figure 16.3, type I error rates are very high too). Basically, the item p-value is flagging *all* items regardless of exposure. This is because the statistics themselves were drifting higher because of the increase in ability. The number of administrations needed for detection were substantially lower for the moving average p-value statistic compared

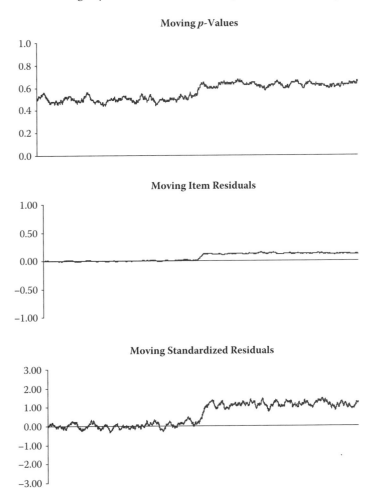

Figure 16.7 Plot of item exposure statistics for item 5 (b = 0.0, a = 0.7). (Normal ability distribution, ρ = 0.25, 100% exposure, k = 100.)

to the other two exposure detection statistics. This is because the p-values were already drifting off to 1.0 because of the shift in distribution and well before the exposure had even been introduced into the simulation. As the cutscores were set under the ρ = 0 case, everything looked acceptable for type I error. But had they been set under this particular set of simulations they would have been unstable and inaccurate. The item p-values were already drifting off to 1.0 before any exposure was introduced.

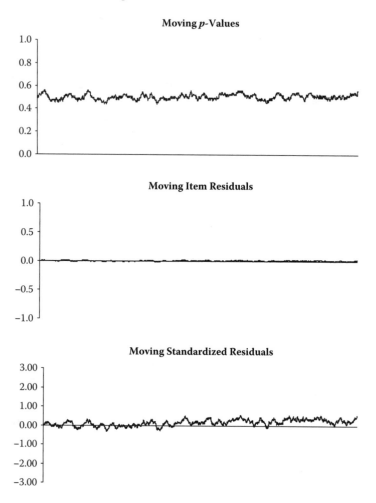

Figure 16.8 Plot of item exposure statistics for item 5 (b = 0.0, a = 0.7). (Normal ability distribution, ρ = 0.25, 10% exposure, k = 100.)

Looking at the results in their totality, and disregarding some of the irregularities and minor trends in the findings, we were struck by the similarity of results for the two IRT-based exposure detection statistics across the three ability distributions compared to the results observed with the moving average p-value statistic. The IRT-based exposure detection statistics functioned much better.

Conclusions and Lessons Learned

The results from the study were revealing for all of the variables studied: (a) ability shifts, (b) item exposure models, (c) item exposure detection statistics, and (d) item statistics. First, the ability shifts were consequential. As a starter, it was easy to see that the moving p-values produced unacceptable results when shifts in the ability distribution took place over the testing window—basically all items would be flagged with shifts in the ability distribution, regardless of whether or not they were exposed. In those situations, clearly, the other two statistics would be preferred. With a constant normal distribution of ability over the testing window all three statistics produced comparable results.

With respect to the item exposure models, putting aside the somewhat unrealistic first case ($\rho = 1$, 100%) where detection was easy, one finding was that the $\rho = .25$, 10% case produced quite unacceptable levels of exposed item detection. This is the case where 10% of the examinees were given a small boost in their performance level because of prior knowledge. For an examinee with a 50% probability of success on an item, that success was increased to 62.5% under the item exposure model. For a more capable examinee with a probability of success of 75%, that success was increased to 81.2%. For an examinee operating at chance level (25% probability of successfully answering an item) that probability would be increased to 43.75%, far from any assurance of a correct response to the item. And in this condition, these increased probabilities would be applied to the item level performance of only 10% of the examinees. Clearly, this level of item exposure would be very difficult to spot in practice. The levels of detection of exposure were substantially higher in the other two cases, but especially so for the case $\rho = .25$ and 100% exposure. How realistic this case might be in practice is not certain, but the detection rates were certainly quite good.

As for the item exposure detection statistics, our research showed a strong advantage to the two IRT-based statistics. They were applicable across all conditions simulated, whereas the item p-value was not. And, they typically identified exposed items except in the cases where a small amount of exposure was simulated. We noticed too, that whatever the detection rate, it was always easiest to detect the more difficult items, and generally the more discriminating items. Some minor reversals were seen in the data however.

Interestingly and importantly, the findings about the item exposure detection statistics and how they functioned are applicable to all forms of computer-based testing from linear or linear-on-the-fly to multistage, to fully adaptive tests (see, for example, van der Linden & Glas, 2000). Once an item is administered in whatever design is operative in the testing program, the examinee performance data can be added to the string of data being collected on each item, and the item detection statistics can be updated, and tested for significance. An item remains in the bank until it is retired or identified as being exposed. The likelihood of detection of exposed items obviously depends on the confidence bands that have been established (which depend on the window size, in this study the number of examinees used in the statistics was 100), the statistical characteristics of the test items, and the type of exposure taking place. For the two IRT-based statistics, that considered examinee ability in the calculations; the nature of the ability distribution was irrelevant. This is an important property of a useful item exposure detection statistic. We are pleased too to discover that the more discriminating items were the ones that could be detected fastest. These are the same items that influence the ability estimates the most, and therefore, when exposed, they contribute more than other items to the invalidity of examinee scores.

We were pleased with the results from the study and expect to continue on with the research. Obviously, we are looking forward to seeing the statistics actually used in practice, which we expect to happen soon. A number of testing agencies are now considering the addition of statistics like those we have reported on in this chapter. Also, next steps in this research probably will focus on just one of the item detection statistics—item standardized residuals—and investigate additional item exposure models. Other detection flags are also possible too. For example, examinee response time on items is being compiled. Were examinees to answer an item correctly using substantially less time than other examinees, a question would be raised about the validity of the examinee's response. Possibly, this information can be combined with the item detection statistic to more rapidly identify exposed items (for important research on the use of item response times in detecting exposed test items, see van der Linden, in press). Clearly there is a lot of work to be done, and this general approach appears to have considerable merit, and would be relatively easy to implement in computer-based testing programs.

References

Ariel, A., van der Linden, W. J., & Veldkamp, B. P. (2006). A strategy for optimizing item pool management. *Journal of Educational Measurement, 43*(1), 85–96.

Hambleton, R. K., Swaminathan, H., & Rogers, H. J. (1991). *Fundamentals of item response theory*. Newbury Park, CA: Sage Publications.

Han, N. (2003). *Using moving averages to assess test and item security in computer-based testing* (Center for Educational Assessment Research Report No. 468). Amherst, MA: University of Massachusetts, School of Education.

Han, N. (2006). *Detecting exposed items in computer-based testing*. Unpublished doctoral dissertation, University of Massachusetts, Amherst, 2006.

Han, N., & Hambleton, R. K. (in press). Using moving averages to detect exposed test items in computer-based testing. In S. Sawilowsky (Ed.), *Real data analysis*. Greenwich, CT: Information Age Publishers.

Impara, J. C., & Foster, D. (2006). Item and test development strategies to minimize test fraud. In S. M. Downing & T. M. Haladyna (Eds.), *Handbook of test development* (pp. 91–114). Mahwah, NJ: Lawrence Erlbaum Associates.

Jodoin, M., Zenisky, A., & Hambleton, R. K. (2006). Comparison of the psychometric properties of several computer-based test designs for credentialing exams with multiple purposes. *Applied Measurement in Education, 19*(3), 203–220.

Karabatsos, G. (2003). Comparing the aberrant response detection performance of thirty-six person-fit statistics. *Applied Measurement in Education, 16*(4), 277–298.

Lu, Y., & Hambleton, R. K. (2004). Statistics for detecting disclosed items in a CAT environment. *Metodologia de Las Ciencias del Comportamiento, 5*(2), 225–242.

Pitoniak, M. (2002). *Automatic item generation methodology in theory and practice* (Center for Educational Assessment Research Report No. 444). Amherst, MA: University of Massachusetts, School of Education.

Revuelta, J., & Ponsoda, V. (1998). A comparison of item exposure control methods in computerized adaptive testing. *Journal of Educational Measurement, 35*, 311–327.

Segall, D. O. (2001, April). *Measuring test compromise in high-stakes computerized adaptive testing: A Bayesian strategy for surrogate test-taker detection*. Paper presented at the meeting of the National Council on Measurement in Education, Seattle, WA.

Stocking, M. L., & Lewis, C. (1998). Controlling item exposure conditional on ability in computerized adaptive testing. *Journal of Educational and Behavioral Statistics, 23*, 57–75.

van der Linden, W. J. (in press). A hierarchical framework for modeling speed and accuracy on test items. *Psychometrika*.

van der Linden, W. J., & Glas, C. (Eds.). (2000). *Computer-adaptive testing: Theory and practice*. Boston: Kluwer Academic Publishers.

Yi, Q., & Chang, H. H. (2003). A-stratified CAT design with content blocking. *British Journal of Mathematical and Statistical Psychology, 56,* 359–378.

Zhu, R., Yu, F., & Liu, S. (2002, April). *Statistical indexes for monitoring item behavior under computer adaptive testing environment*. Paper presented at the meeting of the American Educational Research Association, New Orleans, LA.

17

From Independent to Integrated Systems

Richard Dobbs and Stuart Kahl
Measured Progress, Inc.

Introduction

On February 20, 2003, U.S. Secretary of Education Rod Paige held a meeting in his conference room of over twenty presidents of educational testing companies. When all the participants had settled in, Secretary Paige entered the room and said, "I have just one thing to say—you have to get test results back much faster." With little more said, he turned the meeting over to an undersecretary, and departed.

Increasing Demands

The expression "faster, better, cheaper" has been used on many occasions in the last decade to characterize the increasing demands placed on the testing industry. Indeed, as a result of the passing of the No Child Left Behind Act of 2001 (NCLB), the volume of statewide testing is doubling. Furthermore, because of the accountability provisions of the law, specifically the consequences for schools not meeting their interim targets on the way toward the goal of 100% proficient students by 2014, adequate yearly progress results must be reported in midsummer, even if tests are administered in late spring, as they are in most states.

Even before NCLB, increased emphasis on school accountability was placing greater demands on testing companies, and with the greater capabilities of new technologies, the industry was already responding by developing new systems for processing testing materials and analyzing and reporting results. NCLB has acted as a catalyst, speeding up process improvements.

349

Statewide Testing—Historical Perspective

In their infancy in the late 1970s and early 1980s, statewide accountability testing programs looked nothing like the programs of today.
Some states used off-the-shelf Norm Reference Tests (NRTs). With
low stakes and low visibility of results, the only difference between
this and local usage of the NRTs was who paid for it. Other states
used customized tests. Some early programs were modeled after the
National Assessment of Educational Progress and did not produce
student or even schools' results. (The states sometimes offered to produce student and school results if the schools paid to have more than
a small sample of their students tested.) The early customized programs, which often used different tests every year, typically reported
results several months after the tests were administered. (National
Assessment of Educational Progress took two years at that time.) Of
course, the results almost certainly had little impact in those early
years for many reasons, including the low stakes and visibility and
the lack of student and school-specific information.

    With heightened concerns about school accountability, particu-
larly after the dissemination of the Nation at Risk Report (1983),
many states shifted to census testing, administering their assessments to all students in selected grades. Of course, testing all students
meant that school results, paid for by the taxpayers, were produced
and reported publicly. With somewhat of a scandal in the testing
industry, resulting in part from the states' use of published tests that
were already sitting in cabinets in the schools, more customized programs emerged in the late 1980s and 1990s. This also helped address
security concerns that increased with the stakes. Many of these programs, influenced heavily by curriculum specialists reacting to the
problems associated with the off-the-shelf products, made extensive
use of non-multiple-choice formats requiring human scoring.

    In this middle history of statewide testing, the emphasis of the
programs seemed to be on the effectiveness of school programs,
rather than on the performance of individual students. With testing at selected, nonadjacent grades in many states, school results
fluctuated from year to year because of variations in the groups of
students passing through the tested grades. Thus, it was recommended that multiple years of results be accumulated before schools
draw major conclusions and make important decisions about their
instructional programs. Report turnaround time should not have

been a major concern; however, with all of their students in target grades tested and the turnaround time associated with unchanging NRTs as a model, local educators demanded faster reporting from the customized programs.

The federal government's reauthorization of Title I (Compensatory Education) legislation in the mid-1990s began taking statewide testing to another level. Federal law then required statewide testing in reading and math in one elementary grade, one middle school grade, and one high school grade. The tests had to be aligned with state content standards, and all subpopulations of students had to be assessed. The U.S. Department of Education was slow in enforcing the law, and for a variety of educational and political reasons, the No Child Left Behind Act of 2001 was passed in January of 2002. This law, which was actually the next reauthorization of Title I, ultimately required testing in reading and math in grades 3 through 8 and one high school grade, as well as the testing of science in three grades a few years later. As mentioned previously, the accountability provisions of the law attach serious consequences to the results if schools fail to reach their targets along the way toward the requirement of 100% proficient students by 2014.

Three characteristics of statewide testing today have necessitated the significant changes in the past five to ten years in how testing companies do things. They are:

- the volume of testing;
- the allowable turnaround time for results; and
- the required accountability for all students.

No Child Left Behind has doubled the amount of statewide testing taking place, and to meet the accountability requirements of the law, results from late spring testing must be reported by midsummer. The testing industry has responded to these requirements in two ways. First, there are more full-service, large-scale assessment companies now than there were a few years ago. Some of the companies who specialized in just one component of testing (e.g., logistics, human scoring of constructed responses, etc.) have developed or acquired the capacity to perform all the steps in an assessment program. Also, some testing companies who were not historically involved in statewide testing have entered that market. Second, companies have used the capabilities of new technologies to develop more automated systems for developing tests, handling testing materials, scoring responses, analyzing data, and reporting results.

At Secretary Paige's meeting with testing executives in 2003, it was pointed out that with proper planning and new high-tech systems, activities such as human scoring, scaling and equating of new tests every year, and so forth, could all be done in a very timely manner. The biggest problem with respect to turnaround time is getting clean, complete raw data files. In the early years of low-stakes statewide assessment (and similarly for local district use of published off-the-shelf tests), results were produced using the data from whatever answer documents were returned to the testing company, regardless of the completeness of those sets or the accuracy of coded information about the students. Today, all students enrolled in the schools must be accounted for, and their results must be counted toward the results for the correct subgroups of students. Typically, for many of today's customized programs using new tests every year, final reportable results cannot be computed for any school until the final statewide results are computed. Schools do not all return their testing materials on time. Sometimes stray student answer sheets are found inserted in used test booklets rather than being returned in the separate containers designated for the return of answer documents. Additionally, it is not uncommon for there to be inaccuracies in the coded information on what subgroups students belong to. It can take months to get complete and accurate student data files. At some point, a decision has to be made that the files are complete and accurate enough to proceed with final analysis and reporting of results of record. But, of course, late returns and corrections lead to additional clean-up analysis and reporting—a nightmare for testing companies.

It was pointed out at Secretary Paige's meeting that the only way the shorter turnaround times can be accomplished would be if the student data files, minus the response data from the tests, can be clean and complete well before test data are available. Many states are now implementing student information systems to accomplish just that. With this information the testing companies can know in advance of testing what students in each school they should expect testing materials from and in which subgroups the students belong. In many states, student identification labels are shipped with testing materials so that they can be affixed to the students' answer documents and their results can be merged with the other data on the students in the student information system. This is the right way to go; however, many states' student information systems are in their infancy and completeness and accuracy are still problems for the time being.

Overview of Changes in Systems and Procedures

We closed the previous section on a historical perspective of statewide testing with the mention of student information systems for a reason. One can easily see the roles that information from these systems can play in a testing contractor's work with respect to communication with schools, print orders for testing materials, accounting for materials, and analysis and reporting. Clearly, integrated data systems are essential for timely and accurate work. Interestingly, however, in the not-too-distant past, major steps in the operation of statewide assessment contracts were often accomplished in isolation. Needless to say, unnecessary redundancy and inefficiency characterized these operations. With low volumes, low stakes, and generous timelines, this was not a problem years ago. No testing company would be in business today using such outdated, inefficient systems. In the following sections, we describe the ways things used to be done in various operational areas and how they have evolved over time. The remainder of the chapter focuses on two particular integrated data systems that Measured Progress has developed which have helped the company to emerge as one of the major players in the statewide assessment industry.

Development and Publication

In the early years of customized statewide assessment, some testing companies certainly used more sophisticated systems than others; however, at some point they all used some pretty archaic methods. When Measured Progress's predecessor, Advanced Systems, incorporated in 1983, word processors were just replacing Selectric II typewriters. Our test (item) developers entered the items, printed them, cut the individual items out along with copies of any art that may go with them, sorted them by content category, and taped them onto paper (several per page), to prepare item sets for copying for review by clients and advisory committees. Item documentation was written on the items: answer keys, reporting category codes, field test results, and so forth.

Once items were edited and approved, the items in final item sets were cut out, sorted across forms if multiple forms were used, and arranged on pages in the approximate locations they would appear on the final tests. These cut-and-paste pages were then handed to desktop publishers who often key stroked everything from scratch, producing (after many reviews for proofing and revising) the camera-ready pages of the tests, including embedded directions. Placeholder art was often on the pages, because our printers had to photograph original art and attach the images to the pages themselves. Print vendors photographed these boards, producing negatives that were laid out on larger sheets corresponding to the signatures that would be printed. The sheets of negatives were used to burn plates that ultimately were mounted on the printers, which ran the required numbers of signatures. Of course, collation of signatures, cropping, and so forth followed.

Needless to say, the past twenty-five years has seen continuous incremental improvement in these procedures. Electronic transfer of items meant that editing staff did not have to keystroke items from scratch, although going from common word-processing software to more sophisticated publishing software created the need for considerable tweaking. Electronic transfer of camera-ready pages and forms to the print vendors speeded the production process and, of course, advances in the printing industry were significant as well.

In the mid-1990s, Measured Progress (then still Advanced Systems) developed iRef™, an electronic repository for test items and their associated documentation codes, item statistics, usage history, and so forth. With this tool, item sets could be generated easily for review purposes, with item statistics attached if the review was occurring after field testing. The interface with desktop publishers was greatly facilitated, except that artwork that may have been associated with items was not captured by iRef and still had to be accessed by editors from separate files. iRef was the predecessor to a new item and assessment banking system that is described later in this chapter.

Shipping and Receiving

At the start of a 1980s state assessment contract, the contractor's program manager would obtain a directory from the client state, identifying the school districts, the superintendents, the schools, and the principals in the state. The program manager would then have word-processing merge files prepared so that personalized letters could be sent to superintendents asking them to verify the names and addresses of schools in their districts and the names of the principals. Then personalized letters were sent to principals asking their enrollments in the grades to be assessed later in the school year. The word-processing data files were sent to shipping and receiving personnel (along with school enrollment numbers) so that they could prepare labels to be affixed to cartons of testing materials to be sent to the schools. Packing lists used to assemble the materials for each school were most likely prepared with some kind of spreadsheet software, if not with word-processing merge files. Labels were also prepared for subsequent mailings of reports of test results to superintendents.

When materials were returned from schools, they were counted by hand and the totals of different items returned from each school were entered by hand in tables which had been prepared before the return of materials that showed the quantities of materials shipped to the schools a few weeks earlier. Follow-up calls to schools were an effort to try to track down missing materials. School header sheets were placed on top of scannable answer sheets so that a school identifier could fill one of the fields in each student's raw data record, allowing each student's results to be associated with the correct school.

Again, continued improvements in procedures occurred over the years. These improvements, however, were small and incremental in nature. Data files used by different groups within a company were still separate and maintained independently by each group for many years. Perhaps the advancement that has made the greatest difference in the logistics of large-scale assessment programs is that in the area of barcode technology. With variable barcodes assigned to each document and each student, the capability of tracking materials has been enhanced greatly. The iCore™ system described in a later section illustrates this point.

Constructed-Response Scoring

For this chapter, *scoring* refers to the assigning of numeric scores by human readers to individual student responses to constructed-response questions. It was, and still is, a common practice to have students write out their short essay responses or show their work on math problems in spaces provided in the same scannable answer documents on which they mark their answers to multiple-choice questions. In the not-too-distant past, when a set of response documents from a school was received by the contractor, the set was broken up—that is, they were distributed into many different cartons so that a single table of readers would not score all the responses from a particular school. Thus, there was some degree of randomness in the assignment of answer documents to different reader tables. Readers would bubble in scores they assigned just below the students' writing spaces on the answer documents. If double-scoring was required, the first reader's scores were covered over by opaque Post-it Notes® so that the second reader could not see them. When all the constructed responses were scored, the answer documents were then sent to scanning to capture all the data encoded on the answer documents by students, school personnel, and readers.

Most companies now use image scanning and image scoring of constructed responses. Measured Progress's iScore system was implemented in 1995. When answer documents are scanned, the demographic and multiple-choice data encoded on the answer documents are captured in files that get passed on for data processing and preliminary analysis. Electronic images of students' responses to constructed-response questions are sent to the iScore system, which randomly assigns them to readers working at computer stations. The system can reassign responses to "blind" second readers, as required, and monitor reader consistency with other readers or with scores assigned by experts prior to the large scoring session. Of course, with image scoring, school sets of answer documents are never broken up and never leave the contractor's warehouse. Images can be sent electronically to scoring centers set up anywhere in the country. The computer system captures the scores the readers enter and produces files of constructed-response item scores, which get passed on to data processing where they are merged with the students' selected-response data.

Analysis and Reporting

Analysis and reporting associated with customized statewide testing programs have advanced considerably over the years. Most of the advancements have been the result of both improved self-developed and third-party software. For example, a combination of both allow the psychometric scaling and equating of a series of tests from one year to the next to be accomplished in hours, rather than days and weeks. Although analysis programs can be prepared in advance of the availability of final data files, final analyses often cannot be accomplished until the raw data files are complete and accurate. Therefore, recovering answer documents from all schools' students still poses some challenges. Computer-delivered testing is supposed to help this situation, but the experiences of many states have suggested that because every school has a unique configuration of hardware and software, the time when all students in a state can take tests within a narrow timeframe (necessitated for security reasons) is still a ways off. Thus, the recovery of all paper testing materials remains a challenge for some time to come.

Advances in reporting software have cut the time between analysis and delivery of reports considerably. Years ago, a page from a multipage school report was treated as a separate document. Thus, page 1 for all schools in a state was printed, and then page 2 for all schools, and so on. Collation and binding of school reports were then accomplished by hand. Now reporting software allows intact school reports to be printed, with no need for collation. Many states allow reports to be posted on secure Web sites, eliminating the printing of certain reports altogether.

Progress toward New, Automated, Integrated Systems

As stated previously, using some of the old procedures and technologies described earlier in this chapter, a testing company would not survive in today's educational environment. Consequently, all the major players in the large-scale assessment industry have developed relatively new systems to accomplish major functions associated with statewide accountability programs. Furthermore, because of the interactive nature of work in the various functional areas, these systems are truly integrated, thereby eliminating a great deal of duplication of effort that was characteristic of older assessment programs and procedures.

A large part of our success is due to the use of new systems handling many operations. For example, we have developed our own online contract planning and specification system, into which program managers enter detailed specifications for each functional area. This is the primary tool for communicating contract scope and requirements to the functional groups. It allows for effective contract planning and operational management. In addition, we created the Program Coordination Unit that is responsible for assisting program management with scheduling, coordinating schedule adjustments to yield the master schedule, and monitoring adherence to schedules.

Finally, and of great significance, our psychometric staff has produced standardized computer analysis routines for such purposes as item analysis, IRT equating, differential item functioning analysis, analysis of decision or classification accuracy, and other key processes. Mentioned earlier, these have reduced the time it has taken to complete some analyses from weeks and days to hours. The same is true for key aspects of our reporting software. All of these efforts began to examine ways in which integration across processes could yield multiple benefits and increase capacity for the anticipated simultaneous increases in volume and expectations for quality.

Basic Issues and Approaches Driving the Change from Independent to Integrated Systems

Thus far, we described the environmental factors influencing the need for change, and we have begun to present various efforts that were undertaken to get us there. The remainder of this chapter looks specifically at the changes, primarily spurred by process improvement efforts, involved in moving from independent to integrated systems. A few points to keep in mind as we make this transition are:

- Customized, large-scale testing programs have several attributes that create opportunity for error. These include test design issues, shared responsibility with clients in an iterative item development and review process, the manner in which student responses are captured and scored, the nature of student registration in schools, and the need for individual student score accountability.
- Current models of assessments of student learning include combinations of dichotomously scored items graded electronically and constructed-response items that require human graders. Results from both item types must be combined in generating student scores.
- Unlike a typical manufacturing environment in which products are distributed, student assessment involves a cycle of shipping

materials to the testing sites, retrieval and receipt of both used and unused testing materials (including answer documents and separate test documents in many cases), and after the appropriate processing and data analysis, return of test results for students and schools.

- There is a challenge involved in getting up-to-date information on students to be tested well in advance of, just prior to, and during the test administration.

- The focus of improved processing has been in the area of unified data files, accomplished in conjunction with states' student information systems and involving considerable prework. A variety of tracking reports were developed to assist in checking and using the information obtained from the system and monitored throughout the process.

- A major emphasis placed on development of new systems has been complemented by the review of processes. This involved meeting with process owners and investigating ways to improve systems by breaking down processing into component parts and looking for bottlenecks, critical handoffs, and determining what was not working well.

- Barcode technology has been used to enable extensive tracking capability.

- *Lean manufacturing* and *one-touch* processing were principles used in generating the new systems and processes. The owner of this initiative was determined to apply principles developed in a manufacturing environment whenever they contributed ways to increase efficiency and reduce the potential for human error. Everything from the physical environment to the equipment used and the training provided were planned to streamline the effort, eliminate redundancies, and use automation to our advantage.

Looking at End-to-End Business Intelligence

Obviously, many of the important developments that are discussed here are based on implementation involving new technologies. Before describing those, it is important to note that they originated through the identification and documentation of fundamental business needs, and these were primarily about the acquisition, management, and dissemination of information. In approaching the problem this way, we also ended up dealing with numerous process and measurement issues.

In getting started, it was determined early on that it was vital to
assess how we use information to drive the business, then to outline
how information is managed and utilized as a company asset. It was
equally important to understand both current needs and a vision of
the future as a prerequisite for laying out a roadmap for improvements. Because significant growth was anticipated as part of that
future, and because the expectations for large-scale assessment were
expected to continue to change, it was important to think about how
key decisions would serve our future and provide the flexibility to
leverage existing investments.

The ensuing effort also included identifying several critical success factors, including (a) senior management buy-in for this initiative, (b) development of a data warehouse to bring together all
the disparate information from across the organization, (c) review
of staffing to make sure the right personnel were on-board, and (d)
selection of the appropriate application software and related analysis, forecasting, and management techniques.

The organizational goal driving this effort was to continue to
be a premier client-focused organization, using best-in-class methods to foster improvements. Because the envisioned solutions were
unlikely to immediately provide everything that could be imag-
ined, we needed to start with something very tangible and very
well-defined, and expand from there. One basic approach was to
treat every information request from our internal and external clients as an enterprise-wide business intelligence request. This influenced the thinking to build out a common, extensible, and scalable
platform and architecture, such that every subsequent request
would be that much easier to accommodate. The systems developed
would need to (a) support the ability to solve business problems,
(b) be accompanied by open, standards-based procedures, processes, and policies, (c) utilize architectures that could be adapted
to envisioned future needs, and (d) offer the desired level of stability and service. Cross-departmental teams looked internally at
what was not working well and where critical bottlenecks needed
to be addressed.

Although standard data models are used more widely in other
industries, there are emerging standards in the education industry; however, for those standards to be effective, schools, districts,
states, Measured Progress, and our competitors will have to adopt
those standards. Until that time, we need to support our corporate

needs and strategy. Part of that technological support is to address the needs of today and forecast the needs for the future. Utilizing a standards-based approach we can maintain a level of nimbleness and flexibility that we wouldn't have otherwise. By taking a standards-based approach to data modeling, we have the ability to not only use third-party software, if necessary, but to change or react quickly to industry changes or more simply to customer needs. We loosely follow various standards such as School Interoperability Framework and IMS Global Learning Consortium.

Business Performance Management

In line with the concept of integration across processes, it was critical to give serious consideration to information in the context of performance management, and in particular to look at the transfer and sharing of data, the interdependencies across various processes, and the requisite information flow. This involved understanding the concurrent responses to information requests in multiple areas and ensuring that all are done in alignment with the company's growth. We looked at data flow and process requirements, which required a substantial needs analysis and discussions with process owners across the company.

Another important consideration was ownership of information. Successful new or reengineered processes require a multidisciplinary team whose focus is to understand current needs and to create the vision for the future. With that in mind, we set about taking the necessary incremental steps for making the focus on information and information-sharing a strategic initiative. This alignment played a role in guiding projects, managing requirements, creating standards, and assisting with technology selectio' and implementation. It meant getting the right people at the ta' and included representation from Client Services (our pros managers) and the business units, those with information nd skills and an ability to understand and relate to the pror⁰ed priorities of our business. Many of these processes we·ᵃem. earlier, as were some very successful prior efforts to ;ᵈ the Key group members who understood the proc ally the interrelationships were indentified, and these v ⁿgage in process owners. They understood the work extensive discussion, collective analysis, and targeted redesign of work flow.

Several methodologies and frameworks were considered and—at least in part—utilized to guide these efforts. Among those was the Balanced Scorecard, perhaps the most well-known methodology for developing key performance indicators based on both financial and nonfinancial business objectives. This methodology allows one to think in terms of people, workflow, and capital that contribute to processes and activities, which support customers, who in turn generate revenue and profit. The cause and effect linkages helped us understand the global view so that we balanced each aspect.

Once objectives are established, the next step is to think about targets, or what has to happen in order to achieve the objective. This in turn creates a need for development of metrics that track progress, which has, as described in the pages to follow, turned out to be more difficult than it might have appeared. Another challenge is the immediacy of ongoing project deadlines and finding the time to step outside the demands of ongoing work. Then, of course, there is the fundamental complexity of that work. This was accomplished as part of, and in line with a strategic plan, and the work was defined as the processes were being redesigned.

Process Definitions and Performance Metrics

Regardless of how we define business performance management, metrics and their use in dashboards contribute to how an organization conveys concise and timely information. We realized that there is still much to be learned about metrics, how they should be defined, and how we can determine their relevance and validity. It was also recognized that developing them would be driven by clear delineation and understanding of our processes.

A dictionary definition of *process* is "a series of continuous actions operations concluding to an end." Measured Progress has undertaken several process improvement efforts over its history; and most of which, although successful in generating important changes, were on a departmental basis. Due to the increasing complexity of its efforts as well as the interrelationships among our processes, we expanded to include our acknowledgment of the input and output for a process. This focus on interrelationships was central to the development of new systems and is also proving to be extremely important in their utilization.

In our workflow, processes have many interface points where inter-departmental processes exchange inputs/outputs, that is, where our touch points exist. In Measured Progress relationships—whether with clients, regulatory agencies, vendors, or markets—we see multiple external process interfaces. Some examples of relationships involving internal process interfaces are accounting-to-finance, operations-to-accounting, touch points-to-master schedule, and assessment material-to-distribution. When these relationships were studied, it became obvious that a fair number of processes and interfaces are relevant to this enterprise.

As clearly indicated in the voluminous documentation provided by our process improvement initiatives, within and between processes there are a series of discrete tasks connected by dependencies. Each task, in turn, consists of specific operations or procedures. It is at least theoretically possible to get very specific with respect to measures for each operation, task input, task output, dependency chain, interface point, and resource. The measures taken would, ideally, be the basis for metrics that provide critical information regarding our enterprise performance; however, this is not to say that they all ought to be measured and monitored. Realistically, we may not be able to obtain timely information or cost-effectively report on every measure, so the question, "What do we want to measure and report?" was a topic of much discussion as we approached development of integrated solutions.

Our focus was on interdependencies between processes and within processes, with a deliberate strategy to look at both the front and back ends, starting with the scoring operation and moving to include other critical areas of publications, development, scanning, and research/analysis. Specific reports were developed, for example, to assist program management in monitoring and resolving discrepancies in a timely manner and in a way that prevents errors and saves time.

Steps toward Improved Processes   The evolution of process, procedures, and technology at Measured Progress has been a company-wide effort. Our basic approach to process improvement, procedural changes, and technology adoption was to keep it simple. By trying not to make too many changes at one time we were able to make major improvements in our process and procedures and increase our efficiency. Some of the steps we took to improve our processes are as follows:

1. *Realistic goals*: We made a conscious effort to set realistic strategic goals. It is very easy to get caught up trying to do too much at one time or too soon.

2. *Ownership of plan*: This cannot be emphasized enough. We create a plan and we ensure that people feel a sense of ownership in the plan. When all parties involved in the process feel a sense of ownership, execution of the plan is much easier.

3. *Document or review current process and procedures*: If there is previous documentation, it is reviewed to ensure the team responsible for the process/procedure improvement understands the process. If there is no or minimal documentation, it is updated as well.

4. *Observe*: If possible, we like to watch a process in action. There are times when documentation does not enable one to understand how something is accomplished alone (i.e., "a picture is worth a thousand words").

5. *Communicate with users*: Communication is very important to process improvement. The slightest change to a process must be effectively communicated. The communication includes documentation, training, and a walk-through of the changes in a simulated environment.

6. *Transparent technology*: Technology is needed many times to achieve our goals. We firmly believe technology must be as transparent as possible to end-users. We want our employees to concentrate on the task at hand, not on how to get the technology to work for them.

7. *Build on success*: This simple approach to process and procedure improvement allows us to have a successful implementation and then gradually accomplish more changes, if needed. Our commitment to setting goals, creating a plan, fully understanding processes, communication, and keeping the whole process simple, gives us the opportunity to build on our previous success.

Determining Process-Based Performance Metrics

What also became very clear in the course of process improvement efforts is the value of metrics as measurements in relation to explicitly specified levels of performance. For us, metrics can be derived from the work of research, manufacturing, finance, operations, marketing, or any other process-based activity. Regardless of their number or availability, though, when they are used to measure performance, users view them relative to a stated policy or benchmark. The management team, in concert with key personnel and subject-matter experts, is addressing this by establishing some initial benchmarks, which will be refined over time by using actual measurements and statistics.

Until recently, measures of performance and improvement have been confined to either departmental views of their specific areas or to more global company performance relative to client satisfaction. Recent efforts, however, have revolved around establishing additional metrics related to enterprise-wide operational issues. These will, we would hope, provide a rich source of material for future articles on this subject.

Corporate Metrics

Measured Progress developed corporate metrics as part of a Five-Year Strategic Plan in January 2004. Those metrics most germane to this discussion were in relation to Goal 1 of four strategic goals. It is: "Be the provider of the highest-quality assessment and related professional development products and services in the state and local K–12 education market." Quality is defined in terms of (a) potential to improve teaching and learning, (b) meeting client needs, (c) technical quality, (d) timeliness, and (e) accuracy. The metrics for this goal relate to:

- Research and use-and-impact survey results (e.g., as part of regular testing program questionnaires one time per contract).
- Client and field satisfaction (determined through surveys).
- Technical Advisory Committee support of our designs and procedures.
- Performance with respect to delivery dates.
- Incidence of errors, penalties, and damages.

As mentioned earlier, attention has now turned to the development of a dashboard approach and to the establishment of process-based performance metrics. In conjunction with, and in anticipation of many of these activities, systems were being developed to provide for the collection, analysis, and dissemination of information critical to all of the major company functions previously described. The two systems that are the main focus of the remainder of this chapter were developed and are being implemented by Measured Progress, but they exemplify the kinds of systems many testing companies are using.

Item and Assessment Banking System

A major initiative undertaken in response to increased demands and in line with our strategic goals was the phased development of a comprehensive, Internet-based technology system to both integrate the various elements of the test development process and to ensure the quality of our assessment products and services. The thought behind this effort was that although human error occurs, it can be dramatically reduced through the effective and creative use of technology to gain better ongoing command over the processes of error detection and quality control.

The resulting solution allows disparate and remote access to the system, ensuring that all contiguous-grade test-related and item-related information resides in a single-source data repository. Items are available for internal review as soon as development begins. This information is used by curriculum experts, clients, test constructors, publications staff, as well as by psychometric/analysis/reporting staff. The Measured Progress Item and Assessment Banking System (IABS):

- includes a complete inventory of all item data, including item stems, clusters, graphics, open-response scoring rubrics, multiple-choice options, answer keys, reading passages, and all associated usage characteristics and statistics for every administration;
- has a retrieval interface that makes the items and pilot results accessible to multiple users, including members of client states' staff with remote access capability;
- enables access to item data and statistics via standard reports or as exported tab-delimited files for import to standard spreadsheet or database applications;
- allows users rapid search and retrieval of items and groups of items based on item content characteristics as well as multiple models of item statistics;
- includes a remote highly secure SSL encrypted Web-based interface for multiple authorized users, maintaining security of items with controlled access;
- provides a simple mechanism to recognize the item status and development history, for example, item and assessment usage by administration, form (number and year), release status;

- allows for the capacity to build test forms for all content areas with easy selection of items, swapping of items, and selection of items to target test form characteristics while displaying an indicator of items selected for specific test forms and statistical analysis of test forms for equating purposes.

Put simply, IABS is an integrated enterprise system for the full lifecycle of item and test development. This system offers the secure access, functionality, flexibility, and scalability of a world-class electronic item bank.

In line with our attempt to build systems for now *and* the future, integration in this area is also a key element in moving to increased use of computer-based testing (another way to certainly explore faster, and potentially better, if not cheaper). Beyond the traditional usages of an item banking system, a critical first step in preparing for computer-based administration is preparation of item and test content for the computer. Item conversion for computer-based administration is a complex effort which is highly dependent on the source data. Each item bank conversion effort—and in regard to state-specific or local district initiatives our experience is that there will be numerous such efforts—is a unique Extraction-Transformation-Load process, and is highly dependent on the quality and availability of data in an electronic format. The most complex and time intensive Extraction-Transformation-Load efforts take data from disparate sources (paper and electronic) and store them in one single repository. The simplest conversions are straightforward data transformation services packages that map item data from one database to another. In any event, there needs to be a mechanism for such transformation as well as a means for all item and assessment data to reside in a standard format for easy access and retrieval.

To summarize the benefits, the IABS provides:

- a measurable, repeatable item and test development process that is scalable to meet future contract needs in a cost-effective manner;
- a standards-based data model that allows us to effectively leverage items from third parties;
- efficient storage and retrieval of item and test data that enhances users' ability to find and repurpose content;
- automated processes that reduce opportunities for error and shorten development timelines;
- item versioning and auditing that reduces risk of data loss and corruption;

- a single-source data repository that facilitates the review process, thus increasing efficiency, shortening timelines, and capturing critical decisions during item review. All item-level decisions will be traceable to the event and to the individual making the decision, with provision for decision commentary.

The Measured Progress Operational Enterprise System

As earlier noted, during Secretary Paige's meeting in 2003, it was stated that human scoring, scaling and equating of new tests every year, and so on, could all be done in a very timely manner with proper planning and new high-tech systems. Later, it is mentioned that the biggest problem with respect to turnaround time is getting clean, complete raw data files. Finally, one of the basic issues mentioned as pushing us toward integrated systems was the cycle of shipping, retrieving, scoring and data processing and analysis, and returning scored results. All of these issues, and the related processes and information needs, drove the development of what came to be known as our Operational Enterprise System.

The Operational Enterprise System captures, generates, and retrieves data at all points of work and service in real time. The fully integrated system is designed to drive business performance and revenue by eliminating boundaries of time, distance, and scale; the goals are to achieve the desired quality, increase customer satisfaction, decrease cycle time, increase throughput process, and reduce errors. Hence, it is not only built to handle more volume, but also to ultimately help us achieve "faster, better, and cheaper." Modules include iCore™, iTrack™, ScanQuest™, Distributed iScore™, iServices™, and iReports™. Major functions, including program management, research and analysis, data processing, distribution, login, scanning, and scoring, as well as client information, are integrated on an Internet application.

iCore

The iCore system is a Web-based application, implemented using multitier scalable architecture with a relational database management system at the back end and Web browser-based *thick client* as the user interface. iCore database is a core corporate database that contains current and historical information regarding different business entities, such as county, district, and school with their associated addresses, contacts, phone numbers, and so on. In addition, it contains enrollment, grade, and order information. It is an essential source of information for shipping, project management, research and analysis, data processing, and other groups. The iCore database provides data integrity, data securing, and automatic database maintenance. The *thin client* application allows users to interface with iCore over the Web. iCore interfaces to iTrack and to carrier systems to track assessment materials, and it allows clients to order additional materials online and request a UPS pickup online.

iTrack

iTrack is a dynamic Web-based search engine that tracks, records, accounts, and reports unique serial number bar-codes for all secured test material from print creation (at print vendor) to on-site (in-house) delivery. In addition, iTrack identifies and validates QA/QC process of all materials, controls packing quantities, authenticates secured serial assignment to school/districts, produces packing slip at box level, produces shipment summary report, tracks the material in transit, date/time stamps delivery, monitors and validates returned material, records received material, tracks in-house login (accountability of all secured material), records edits, assigns material for scanning, records location of material in storage, and produces real time discrepancy reports.

By integrating with iCore, all contract, client, and order details are entered once. This eliminates errors and redundancy and yields consistency of data by creating a centralized source for housing this information. Identification, tracking, and reconciliation of secured materials are facilitated through:

- *Packing*: Tracking numbers for each secured document are linked to individual shipping locations. These numbers are captured systematically and identified on detailed packing slips in every box shipped. Each box is also linked to the shipping location for further traceability.
- *Receiving*: The first step of reconciliation, iTrack links each box received to the location where it was shipped. This allows for a "big picture" look at materials shipped versus materials received.

- *Login*: The next step in reconciliation of materials. Used answer documents are systemically received for further internal processing, and returned secure materials are reconciled by location to identify any outstanding items.
- *Internal tracking of inventory and materials*: Linking all received materials to in-house tracking numbers allows for tracking of material through our process. By utilizing static pallet numbers and location identifiers, inventory can be found systemically once assigned to a specific location in our warehouse.
- *Status reporting*: Utilizing all of the previously mentioned features, iTrack can generate daily status reports of materials for each contract. This allows for real time updates for internal management and provides important information to our clients.

ScanQuest

Once testing documents are used and returned, a critical component of the operation is the accurate capture of student responses to both selected-response and constructed-response items. ScanQuest is a patent-pending software/hardware system designed for the imaging of assessment material such that it handles the image capture, extraction of data, clipping of responses to short or extended-response questions, and scoring of test/answer/integrated booklets that are returned from a test administration.

The ScanQuest solution provides the scalable capture subsystem with centralized data and file system repository to house Optical Mark Recognition/Optical Character Recognition/Intelligent Character Recognition (OMR/OCR/ICR) data and electronically captured images. The system is used internally by the scanning department to electronically capture paper assessment booklets and provide scanning metrics (reports) such as batches scanned, volume, throughput, operation efficiency, and percentage scan completed. It provides data sets for data processing and for Distributed iScore, as well as secured system reporting to internal customers via the Web.

ScanQuest provides many benefits supporting "better, faster, and cheaper," including:

- integration with the Operational Enterprise System for real time reporting;
- automatic loading of imaged clips into the Distributed iScore system;
- accurate (100%) detection of blank responses;
- utilization of OMR/OCR/ICR for complete document integrity;

- online (browser-based) viewing of answer documents, thus eliminating physical edit pulls, and the ability to remotely manage tasks and view reports;
- functionality for image capture management, which archives and delivers the complete imaged answer booklet across the secured enterprise for online viewing via intranet/Internet, with images provided in color JPEG and Tiff formats;
- elimination of ties to proprietary printing requirements.

Distributed iScore

This system is a Web-based application, implemented using a multi-tiered, scalable architecture. It was developed to allow qualified personnel to read and score student responses from multiple locations at any time. iScore is a *thin client* solution that allows users quick and easy access, and provides these benefits:

- *Increased efficiency*: Distributed iScore has streamlined many processes that were manually done in the past and added other efficiencies to current processes, resulting in labor savings and increased speed of the system.
- *Enterprise integration*: Because it is fully integrated with other enterprise systems, data and images from ScanQuest, contract information for iCore, interaction with iTrack, and passing data to Research & Analysis are all part of the Distributed iScore enterprise integration.
- *Flexible response formats*: Responses can be viewed in JPEG format, and the system accepts XML data to display to readers, providing flexibility with browsers on Windows-based, as well as Macintosh computers.
- *Caching of responses*: The system architecture allows us to cache responses for each reader without writing any information to the reader's computer. Caching allows us to have the next response ready for the readers and maintain our expected levels of system performance. This approach also allows us to maintain the level of security necessary for distributed scoring.

iServices and iReports

iServices is a suite of applications available for use by internal and external clients for a variety of activities. The following service functions have been developed:

- *iRegister™ for Student Registration*: This service is vital to "faster, better, and cheaper" because it addresses the issue of accurate and up-to-date information. It collects contract specific information over the Web prior to test administration. With enrollment numbers, demographics, and student identification information being collected months prior to the test administration for the preslugging of test material, a need existed for the registration of transferring students during this timeframe. The student registration module provides the capability of online individual registration, during the period after the schools receive the preslugged test/answer documents and up to the day of testing.
- *iEnrollment™*: Captures school and district contact information, shipping addresses, product quantities, and enrollment counts.
  UPS pickup: enables requests for material pickup by UPS.
  Additional materials request: Allows users to request additional materials online for upcoming testing events.
  Student data verification: Allows viewing and editing of student demographic data online. Data is received from iCore, either from the state, district, or school and is subsequently passed to the state, school and district administrators, program management, the Program Coordination Unit, and Research and Analysis.
  Integration: Data is received from iCore as well as iRegister and is linked to iTrack.

Early Benefits of the Operational Enterprise System Approach

Comparative annual improvements have been tracked throughout the years during the development and early implementation of the Operational Enterprise System. A few examples of summary data indicating progress toward "faster, better, and cheaper" follow:

Distribution: Using iCore, iTrack, iLogin, and One Touch Login, shipping errors were dramatically reduced between 2000 and 2003 as shown in Table 17.1. The login throughput process increased 367% from 2000 to 2003, and the temporary labor cost per student attributable to login activities was reduced by 72% between the 2000–2001 and 2002–2003 school years.

Scanning: Utilizing ScanQuest, the rate of scanned images increased 326% from 2001 to 2003. The ability to acquire nonproprietary printing has produced an average cost savings of 31.8% in bids for contracts, and savings in scoring costs are estimated to be from 8% to 12%. Conversion errors have been eliminated, and the

TABLE 17.1   Shipping Error Rates from 2000 to 2003

Year	Total Shipments	Shipping Errors	Error % Rate	Sigma Level
2000	27,577	232	0.84	3.89
2001	41,366	143	0.35	4.21
2002	62,049	12	0.02	5.05
2003	54,942	3	0.0072	5.37

elimination of edit pulls produced substantial savings during the 2004–2005 school year.

iServices: Although we are still in the process of refining the metrics associated with these activities, the enhanced communication with clients created by real time data has contributed to numerous indications of increased customer satisfaction. In addition, there are anticipated reductions in mailing costs, labor costs, and time required to complete many formerly manual tasks.

Conclusions and Lessons Learned

Moving from independent to integrated systems has been an exciting time of high expectations, rapid learning, and tremendous collaboration. There are still challenges, and these will come as no surprise. New systems require new ways of working, and because people are still so vital to the processes, getting people involved, trained, and supported is critical to the success of such implementation. Managing timelines and transitions, communicating effectively, and getting folks to actually use the new systems are things not to be assumed. And, by the way, we've yet to see a new system become fully operational without some glitches and fixes along the way.

Meanwhile, there is no halt, or even slow-down, to the work at hand. The addition of more grade levels and content areas means substantial item and test development on a yearly basis and a major increase in demand for scoring services. Standard setting for those grade levels and content areas not previously assessed, requests for additional (and new types of) reports, and increasing pressure on schools to achieve adequate yearly progress goals have the spotlight shining even more brightly on large-scale assessment providers. Ongoing operations cannot be suspended in order to take time to review processes, implement new systems, and achieve "faster, better, cheaper." In fact, all of this must be done simultaneously. That being said, the benefits of going this route are clear and compelling. In order to come closer to fully realizing those benefits, another overall look at the end-to-end business processes is underway to make sure that our systems and processes are well-understood and fully implemented across the organization. In addition, the notion of Business Change Management has been introduced to enable us to make continuous, orderly and, as we look to the future, effective improvements to what are likely to be increasingly integrated systems and processes.

References

The Commission on Excellence in Education. (1983). *A nation at risk: The imperative for educational reform*. Washington, DC: Author. Retrieved May 23, 2007, from http://www.ed.gov/pubs/NatAtRisk/title.html

Part VI

The Future of Quality in the Testing Industry

18

Next Steps in Improving Quality in Testing

Cheryl L. Wild
Wild & Associates, Inc.

Rohit Ramaswamy
Oriel Incorporated

Introduction

The discipline of process management has grown from its infancy in quality assurance tools developed in the Bell System in the early 1900s to the complex choice of process management tools available today. Few managers in the testing industry have escaped discussions of Six Sigma, lean manufacturing, Baldridge Awards, the Deming Prize, Balanced Scorecard, project management, ISO 9000, quality improvement, quality planning, quality management, reengineering, and benchmarking.

In a study of best practice organizations that encompassed all the process tools listed previously into a euphuism called business process management (BPM), the American Productivity and Quality Center (APQC, 2005) found that

> Most of the primary benefits of BPM for the best-practice partners can be traced to strategic alignment from top leadership to the front line and the customer, as well as the rigor behind their process improvement efforts. In addition, partners installed processes that involve relentlessly measuring, tracking, and reporting performance that is aligned with business outcomes. (p. 25)

The report goes on to say that although different organizations may take different paths in initiating BPM efforts, each involves some type of enterprise-wide initiative led by the CEO and top officers, business process redesign, and business process improvement.

Improving Testing is organized around the idea of using process tools to improve the business of testing. As described in chapter 2, an effective process management effort involves engaged leadership, conformance to standards (both internal and external), effective design and planning of processes, and continuous improvement of processes. This framework is consistent with what APQC found to be components of BPM efforts of best practice organizations. The diverse authors in this book described how process tools may be applied in the testing industry and presented examples of their application within the industry.

Unlike the organizations in the APQC study, the testing industry as a whole is just beginning to use business process management. Testing organizations typically have several of the ingredients of the quality framework embedded in their normal operations. Quality control and inspection, some application of standards, some level of documentation of operating procedures, and use of technology to manage and deliver testing services are in use in most testing organizations. These practices, however, are neither routine nor systematic in most testing organizations, and therefore the benefits of these methodologies are not often fully realized. Organizations that want to truly differentiate themselves by reducing errors in their testing processes must design and implement a program that deploys all the ingredients described in this book in a systematic, disciplined, and sustained way.

The next section of this chapter describes how organizations can implement an enterprise-wide program of process-driven quality using all components of the quality framework. Based on the assumptions that many testing organizations are in the process of implementing or will implement a process-based quality management organization, this chapter predicts how the testing industry will change as a result of the changing emphasis on quality.

Implementing Enterprise-Wide Quality Programs

A quality program can be thought of as the recipe that defines the sequence and quantities of individual quality improvement activities to be combined to ensure consistent and excellent quality of testing processes over time. Any organization, big or small, that is on the quality journey needs a quality improvement program. For larger organizations, this program may involve multiple projects, a large number of resources, several years to complete, and a significant technology investment. Smaller organizations may implement a more limited program, with fewer projects and resources and a time frame of a few months. But the difference between large and small organizations is only one of scale, not of substance. The same approach can be applied irrespective of organizational size.

As mentioned previously, the quality program outlines the sequence in which improvement activities are implemented to ensure a successful and sustainable deployment. A quality deployment program for a small or large organization involves following ten basic steps.

Step 1: Engage Leadership Team and Key Stakeholders

In chapter 3, Richard Smith wrote about the critical success factors for successful improvement initiatives. Commitment of the CEO and senior staff is critical to the success of a quality initiative. How can this be achieved? A CEO may want to do an organizational assessment of the organization's quality management system. Often leaders have assumed that their management system is best in class until outside eyes report on findings and/or gaps. Leadership retreats involving training and/or workshops with the leadership team to help everyone understand the quality initiative are also useful ways to obtain buy-in. Ultimately, in order to show commitment and engagement, the leadership team must commit organizational resources, money, and time to the quality initiative.

Step 2: Identify Improvement Areas

Where are improvements strategically or critically needed in the organization? These can be identified based on customer complaints, market share erosion, costs of poor quality, and internal or external audits. It is important to focus on the most critical needs and not try to solve all problems at once. Focusing on critical needs emphasizes the importance of the initiative, enables the long-term engagement of leadership (leaders focus on critical issues within an organization), and ensures continued financial support for the initiative.

Step 3: Launch Early Pilot Projects

Successful pilot projects help demonstrate the usefulness of the quality initiative, not only to the team members and the leadership, but also to the rest of the employees in the organization. These pilot projects provide stories and concrete examples of the impact of the quality initiative on how work is conducted, benefits that can be obtained, and so on. They provide lessons learned to the organization—suggesting proactive steps to improve likelihood of success of the next round of projects. In one testing company, the initial pilot projects were quite successful, but they took a long time to complete. The evaluation of the first round of pilots identified two areas for improvement. First, team members were often assigned pilot projects on top of their ongoing work. The recommendation was that team members' work assignments should be adjusted to account for the additional work. Also, teams identified the lack of project management skills as another barrier to success.

Step 4: Implement Ongoing Communication

Communication concerning the quality program is critical. At this stage, a great deal of information about the initial pilot projects is available. Employees want to understand why the quality initiative is being implemented and how it might impact them. Eventually, there will be information about customers' needs and evaluation of quality to communicate. In order to assure effective communication, a communication plan can be developed that answers the following questions:

- What (both key messages and facts) should be communicated?
- Who is the target audience for each communication?
- Who should deliver the communication?
- What media will be used to deliver the message (news release, article, face-to-face communication, etc.)?
- When will the communication occur?
- How will the communication be evaluated?
- How will the communicators be provided with the appropriate information to be communicated?

An example of a communication program at Educational Testing Service is available in Wild, Horney, and Koonce (1996). Successful communication efforts typically use a combination of methods (face-to-face meetings, ongoing newsletters or e-zines, Web postings, etc.) and multiple communicators. Leaders, middle managers, and line managers delivering consistent messages and repeating the messages over time in multiple media is the ideal implementation of a communication plan.

Step 5: Collect Customer Information

Institute a voice of the customer process to identify the quality attributes that matter most to the customer. Voice of the customer is briefly discussed in chapter 14. Understanding what quality attributes are most important to customers and how well customer needs are met provides important information to identifying priorities in selecting projects.

Step 6: Document Key Processes and Collect Metrics

Document the key processes in the organization and implement metrics to assess current performance. The process for documenting processes is described in detail in chapter 13. Process metrics are indicators of performance of the process steps and are indicators of the quality of the process output. As processes are defined, process owners and a process management team may be named to manage the performance of key processes.

Step 7: Launch Next Round of Projects

Use the data from steps 5 and 6 to launch the next round of quality projects. Leadership engagement in the selection of these projects is important and a natural outcome of participation in the monitoring and management of process metrics. The communication plan will include communication about the next round of projects and why they are selected.

Step 8: Monitor Performance Regularly

Performance metrics for processes were identified in step 6 and information was collected about the current performance of processes.

Just like a person's body temperature is not static, the performance of a process is not static. Each process will have a process management team to review and monitor metrics. In addition to the monitoring by the process management team, results for all processes are typically reported and reviewed periodically at the senior management level.

Step 9: Collect Customers' Perceptions of Quality Regularly

Customers' perceptions of quality change over time. The ultimate reason for a quality improvement initiative is to improve the customers' perceptions of quality and service. If an organization doesn't measure customers' perceptions regularly, there is no way to objectively judge the effectiveness of the improvement program. Customers' perceptions of quality can also be used as input into planning for future projects and as measures of outcomes of prior projects. Sharing results of customers' perceptions of quality with staff can help fight complacency and emphasize the importance of customers.

Step 10: Fine-Tune and Reiterate the Process

The quality initiative becomes part of the ongoing business management process. Testing organizations develop annual budgets, go through annual planning initiatives, and (we suggest) review their quality initiative annually. Steps 4 through 9 should be repeated.

The Future for Improving Testing

The authors in *Improving Testing* provided some best-in-class case studies for how quality and process tools can improve the quality and delivery of testing services. These authors are pioneers in the quality movement in testing. As the first Sooners who settled the Oklahoma territory may not have been able to envision the state of Oklahoma as it is today, the descriptions of the use of quality tools in the testing industry may not provide a clear understanding of the implications of these tools to the future of testing. The Oklahoma of today was largely influenced by various environmental factors—including violent weather (tornadoes), underground oil, and a landscape that includes a combination of plains, mountains, and woodlands.

The future of process tools in the testing industry in the next 10 years will also be influenced by the environment surrounding the testing industry. The environmental factors that will be involved are the movement toward computer delivery of tests, the increasing use of test scores for high-stakes decision making, the increasing realization of the need for compulsory use of standards as part of the business of testing, and the internationalization of the testing industry.

One of the major changes in the testing environment has been the move to computer-based testing. Although admissions tests (including the Graduate Record Examinations and the Graduate Management Admissions Test) were in the forefront in adopting computer-based testing, certification examinations have not been far behind. Computer-based testing in the grade school and high school market is just beginning. As issues of availability of technology in the schools are resolved, the use of computers for testing in elementary and secondary education will increase because of the decreased cycle time for providing feedback on scores. Changing from paper-and-pencil testing to computer-based testing involves more than simply copying test questions onto the computer. A new platform entails a whole new way of doing work. In other words, the processes supporting computer-based testing must be invented! There are new software interface issues. Will the test questions transfer appropriately from the item banking software to the test delivery software? In what order will items be delivered and how will the publisher assure that the order does not unfairly influence the test scores? If scores are to be reported immediately, how will the scoring key be checked? Will equating occur before the test administration or after the test administration? How will security of the online test and examinee information be maintained? In summary, this new platform will have many process implications.

A second testing environment factor is the increasing use of test scores for making high-stakes decisions. For example, the No Child Left Behind legislation has raised the stakes on testing in the schools. If students in school districts do not meet state standards, districts are held accountable. In some cases, students may not be promoted or graduate if they do not perform adequately on tests. Passing tests in order to obtain a license are often a prerequisite to employment. The higher stakes increase the motivation for cheating and require new processes to be put in place to prevent or combat cheating. In some cases, the standards for how tests are developed may be more stringent for high-stakes tests than for tests with fewer consequences. (For example, classroom quizzes may be part of the learning process. Although part of a grade, students have the opportunity to discuss the questions with the teacher and the teacher has the opportunity to change the grading criteria when appropriate. The actual discussion of the question may be part of the learning experience. In high-stakes testing, students often never learn the intended key and rarely have the opportunity to discuss the ambiguity of the question with its author. Although examinees may provide comments concerning the question and the publisher may change the key, the opportunity for discussion and improvement is much more limited. The standards for piloting and refining a high-stakes test question prior to operational use should be higher than for a question in an in-class quiz.)

Public awareness about quality problems in testing is a third environmental issue impacting the future of process tools in the industry. Concerns about how to monitor the testing industry are again becoming prominent. Senator LaValle of New York filed state legislation in May of 2006 to create a state board to oversee standardized testing (Arenson, 2006). Toch (2006) recommends creation of an independent testing oversight agency to audit state testing programs and the testing industry. In the fall of 2006, *Educational Measurement: Issues and Practice* published a special issue on standards in testing. In her overview of the issue, Judith Koenig (2006) notes that all seven articles in the issue "point to concerns about compliance" (p. 21) and that there is "renewed interest in enforcement mechanisms" (p. 19). In certification testing, organizations are beginning to use accreditation as a mechanism of monitoring the quality of testing. In chapter 7, Goldsmith and Rosenfeld report that the U.S. Department of Defense is requiring selected employees to be certified by bodies that have been accredited by a third party and that the Conference for Food Protection requires third-party accreditation of food protection manager certification programs.

A fourth environmental factor is the international influence on the use of process tools in the testing industry. As indicated in earlier chapters in this book, standards for testing differ considerably internationally. Generally speaking, more stringent psychometric guidelines are accepted in the United States, whereas third-party review, quality management systems, and security standards are more frequently used internationally. In the certification arena, the desire to have certification recognized is driving international bodies to work cooperatively to standardize how the ISO/IEC 17024 (2003) standard is interpreted. International testing standards are also being driven by the European division of the Association of Test Publishers as described by Marten Roorda in chapter 8.

A final environmental factor, more relevant to the business community as a whole, is the acceptance of the effectiveness of process tools beyond the manufacturing and service industries. The Baldridge Award now has categories for education and for not-for-profits. Several influential quality not-for-profit membership organizations, including the American Society for Quality (ASQ) and the APQC, are working to use process tools in the schools. Testing organizations, whether for-profit or not-for-profit, are businesses and their leaders are aware of quality tools. As evidenced by the chapters in this book, process tools are currently being used in some testing organizations. The stage for the use of process tools has been set.

As the Oklahoma of today may not have been totally predictable to the Sooners of yore, the future of process management and quality in the testing industry may not be clear to the reader. The following presents our vision of the future of process management in the testing industry.

First, we predict the enterprise-wide application of process management tools. Testing organizations are already discovering that in order to compete with the organizations that have been using process management tools, they too must implement effective and efficient processes. Several testing organizations now have process improvement efforts underway, usually focusing on one part of the organization, for example, operations or information technology, rather than the whole organization. Leadership will recognize the need to align the quality effort with enterprise planning, and the quality efforts will soon become organization wide rather than focusing on a single function or product area. In the future, process orientation will move from being a competitive advantage, as it is now, to being a business necessity. Process-oriented testing organizations will develop internal process metrics to monitor and continuously improve their processes.

Second, we predict the testing industry will create industry benchmarks. As testing organizations become more involved in improving processes, they will develop internal metrics as described by Dobbs and Kahl in chapter 17. These internal measures will help to focus and drive internal improvements. As organizations become more self-aware, they also become increasingly interested in comparing their performance to others in the industry. In addition, customer concern about quality of the testing industry will create the demand for benchmarks to better plan and monitor vendor performance. Industry metrics allow an industry to identify quality issues across organizations and address these issues from an industry perspective. Organizations like the Association of Test Publishers and the National Organization for Competency Assurance (NOCA) are likely to become active in collecting such benchmarks.

Third, we predict that adherence to standards will become compulsory. Whether required by government or test sponsors, or pushed by best-in-class testing organizations, third-party verification (whether called certification, accreditation, or attestation) that the testing organization meets quality management, security, data and/or psychometric standards will become commonplace. Chapter 4 reviews the standards and third-party verifications that are currently being used voluntarily in the testing industry. In Chapter 8, Roorda states that "in the United States it is probably not feasible to link review systems to standards. The industry would oppose this vehemently." Mr. Roorda correctly assesses the attitude of the industry currently. We predict that the environment is changing and will provide pressure for change. As described earlier, even the psychometric journals are discussing the need for enforcing standards. As industry leaders recognize the possible advantages to enforcement of collaboratively derived standards, business needs may overcome the current defensive posture of some organizations.

Testing providers often argue that they cannot afford to meet higher standards because their clients choose the lowest price provider and the testing provider must be competitive or go out of business. Test sponsors often argue that they do not know what important quality and product standards are or how to evaluate providers on the standards. By using consistent standards (and standards that are agreed upon jointly by industry and customer groups), the possibility of providing low prices by cutting quality will be diminished. Customers will be provided assurance by third-party auditors that quality standards are being met and not have to make independent decisions on minimum quality standards.

A similar transition has happened in the air-conditioning industry. Energy-efficient air conditioners were more costly than their less energy-efficient cousins and were less likely to be purchased by the consumer. Manufacturers focused their resources on the less efficient air conditioner—that is, the product that paid the bills; however, once energy-efficiency standards were required for all new air conditioners, manufacturers began focusing their design and manufacturing resources on the new models. By removing the lower cost energy-inefficient air conditioners from the market, manufacturers could focus their resources on improving quality in energy-efficient models.

Fourth, we predict that processes to ensure security will be built into the testing endeavor and that standards will be developed to include more detailed security requirements. Security in the testing industry has two aspects—assuring that the tests and questions are secure and that the information about candidates (names, addresses, scores, credit card numbers, phone numbers, and sometimes social security numbers) is secure. Test and item security concerns are related to higher-stakes tests. As tests become higher stakes, the incentive for cheating increases. As paper-and-pencil tests are shifted to computer-based tests, the mechanisms for cheating change and new procedures need to be designed. Chapters 15 and 16 discuss metrics that can be used to monitor the testing process for security breeches. Security metrics and procedures will soon become a standard part of the testing process.

In addition to the processes, standards for security in the testing industry should be adopted and expanded. Will the industry adopt ISO/IEC 27001:2005 or expand security components within other standards? A move is already underway to expand guidance for security for ISO/IEC 17024 and for the *Guidelines for Computer Based Tests* (Association of Test Publishers, 2004). However, ISO/IEC 27001 already has an external review option and has been in use in an earlier version in Britain. Industry-wide discussion will be required to determine the most appropriate path.

Fifth, we predict that testing sponsors will be partners with testing vendors in applying quality tools in the testing industry. In chapter 11, Turner described the importance of vendor relationships in assuring the quality of a testing program. He emphasized that a knowledgeable, well-functioning assessment division in the department of education was a necessary component of effective test delivery. DePascale (chapter 10) also emphasizes the need for assessment divisions to be able to communicate needs to their vendors. Although these two chapters focus on educational testing, certification testing also currently suffers when certifying bodies do not understand certification testing and/or do not have well-designed processes and procedures to deal with their testing vendors. Improvement teams in the testing industry will include both vendors and sponsors in order to maximize improvements.

Finally, we predict that interoperability standards for testing data will be implemented to prevent the opportunity of error as data moves between users and vendors. Both chapters 10 (DePascale) and 17 (Dobbs and Kahl) discuss the need to move data from one computer program to another, from one vendor to another, and/or from one user to another. Seamless interoperability between programs and vendors would prevent the possibility of many errors in the testing industry. Although not possible today, the IMS Question and Test Interoperability Specifications (IMS Global Learning Consortium, Inc., 2006) are a first attempt at moving toward interoperability. Process design and improvement tools will be helpful in designing processes that assure interoperability.

Conclusions and Lessons Learned

Improving Quality is coming to a close, yet the implementation of process quality in the testing industry is just beginning. The testing industry is at a tipping point. To most in the testing industry, quality in testing has been synonymous to psychometric quality. Psychometric quality is necessary but not sufficient to meeting the needs of our customers. Leading-edge testing organizations have discovered process tools and are using these tools as a competitive advantage. Testing sponsors are learning that they are part of the quality equation. The authors in this book have provided a framework and examples of how process and quality tools can be used by testing sponsors and testing providers to improve testing. We hope these examples and ideas improve the quality of testing for your customers.

References

American Productivity and Quality Center. (2005). *Business process management: A consortium benchmarking study best practice report.* Houston, TX: Author.

Arenson, K. W. (2006, May 20). Senator proposes creating board to oversee college admissions tests. *New York Times.* Available online at http://www.nytimes.com/2006/05/20/nyregion/20sat.html?r=1&oref=slogin&pagewanted

Association of Test Publishers. (2004). *Guidelines for computer-based testing.* York, PA: Author. Available at www.testpublishers.org/documents.htm

IMS Global Learning Consortium, Inc. (2006). *IMS question and test interoperability overview: Version 2.1 public draft specification.* Downloaded April 14, 2006, from http://www.imsglobal.org/question/qti_v2plpd/imsqti_oviewv2plpd.html

International Organization for Standardization. (2003). *International standard ISO/IEC 17024 conformity assessment—general requirements for bodies operating certification of persons.* Geneva, Switzerland: Author.

International Organization for Standardization. (2005). *International standard ISO/IEC 27001 information technology—security techniques—information security management systems—requirements (2005-10-15).* Geneva, Switzerland: Author.

Koenig, J. (2006). Introduction and overview: Considering compliance, enforcement, and revisions. *Educational Measurement: Issues and Practice, 25*(3), 18–21.

Toch, T. (2006). *Margins of error: The education testing industry in the no child left behind era.* Retrieved February 11, 2006, from www.educationsector.org/usr_doc/Margins_of_Error.pdf

Wild, C. L., Horney, N., & Koonce, R. (1996, December). Cascading communications creates momentum for change. *HR Magazine,* 94–100.

Contributors

Noel Albertson is the director of Systems, Compliance and Project Management for the CPA Exam at the American Institute of Certified Public Accountants. He was the project director for the CPA Exam's conversion to computer-based testing, which launched in April 2004. He has more than 20 years, experience in information systems. He is an expert in project management and risk management with information systems development, and has developed several methodologies in both areas. His particular passion is identifying and optimizing those factors that lead to successful projects.

David O. Anderson, PhD, is a senior psychometrician and psychometric manager in the Center for Statistical Analysis, Research and Development Division, Educational Testing Service. In this management role, he coordinates operational work and research for the AP, CLEP, ICT, MAPP, and NBPTS testing programs, ensuring the accurate, insightful, and timely production of all client and customer deliverables. Dr. Anderson previously coordinated the operations of the Praxis and GMAT statistical analysis teams, as well as served as a senior business process consultant in the Office of Operational Excellence at ETS. Dr. Anderson received his MA and PhD in educational psychology from Northwestern University.

Roger L. Brauer, PhD, CSP, PE, has served on the board or staff of the Board of Certified Safety Professionals since 1984 and has been the executive director since 1995. Previous to his becoming a member of the BCSP staff in 1990, for 20 years he was a principal investigator at the U.S. Army Construction Engineering Research Laboratory in Champaign, Illinois, and was an adjunct professor at the University of Illinois. He is the author of several books and many publications, holds two patents, and is an inductee of the Safety and Health Hall of Fame International and a Fellow of the American Society of Safety Engineers.

Charles A. DePascale, PhD, is a senior associate at the National Center for the Improvement of Educational Assessment, Inc., in Dover, New Hampshire. His view of large-scale, K through 12 assessment is shaped by experiences as a contractor with Advanced Systems in Measurement and Evaluation, as a client with the Massachusetts Department of Education, as a test-user while serving as a district assessment director and high school mathematics teacher, and as a parent receiving annual score reports for norm-referenced tests and state assessments.

Richard Dobbs is vice president of Client Services and Marketing at Measured Progress, Inc., and is responsible for client services. He also directs the conceptualization, planning, and implementation of product and business development strategies, with particular attention to the integration of standards, assessment, and professional development. Dobbs was previously the vice president of Sales and Marketing for CTB/McGraw-Hill. He has served as an educational consultant to 13 Midwestern states, provided direct educational sales and support to school districts, and served as a classroom teacher. Dobbs holds degrees in elementary education from Western Illinois University and education administration from Northwestern University, and has been a member of the educational testing industry since 1978.

David Foster, PhD, a psychologist and psychometrician, is credited with introducing computerized adaptive testing and simulation-based performance testing as part of Novell's pioneering IT certification program. Since leaving Novell, he cofounded Galton Technologies and later Caveon Test Security. In 2006 he helped create Kryterion where he serves as CEO. A past president of the Association of Test Publishers (ATP), he is active in various advisory boards including the International Test Commission (ITC) and the personnel of the Certification Accreditation Committee and Board of Directors for the American National Standards Institute. Foster holds a PhD in experimental psychology from Brigham Young University.

Sharon M. Goldsmith, PhD, was 2003 recipient of a Distinguished Service Award from the American National Standards Institute (ANSI) for leadership in international standards, accreditation and certification. She serves on several national boards including the

American National Standards Institute (ANSI) Personnel Certification Accreditation Committee and the International Commission on Healthcare Professions Standards Committee. She is chair of the Conference for Food Protection–ANSI Accreditation Committee and former cochair of the Joint Committee on Testing Practices. Dr. Goldsmith directs the Certification, Education and Accreditation Practice of the Plexus Consulting Group, an international firm based in Washington, DC, and is president of Goldsmith International SP.

Ning Han, EdD, was an associate measurement statistician in the Research & Development Division at Educational Testing Service in Princeton, New Jersey. He received his bachelor's degree from East China Normal University of China. He had been a research scientist at the National Education Examination Authority of China before he completed his doctoral studies at the University of Massachusetts at Amherst in 2006. His research interests are in the areas of equating methods, IRT model fit, and detection of exposed test items in computer-based testing environments. Recently, he returned to China to take up a position again with the National Education Examination Authority.

Ronald K. Hambleton, PhD, holds the title of Distinguished University Professor and is chairperson of the Research and Evaluation Methods Program and executive director of the Center for Educational Assessment at the University of Massachusetts. He earned a BA in 1966 from the University of Waterloo with majors in mathematics and psychology, and an MA in 1968 and PhD in 1969 from the University of Toronto with specialties in psychometric methods and statistics. Hambleton has been teaching graduate-level courses in educational and psychological testing, item response theory and applications, classical test theory models and methods, and has offered seminar courses on applied measurement topics at the University of Massachusetts since 1969. His research interests are in the areas of large-scale assessment and item response theory.

Bob Hunt, JD, PhD, manages technical certification testing programs at Cisco Systems, Inc. Prior to Cisco, he served as vice president of legal services at Caveon Test Security and contributed to the formation of Kryterion, Inc., an Internet-based test delivery provider. As director of exam and program management at Certiport, Inc., he designed and developed performance-based certification exams for

Mcrosoft Corp. Dr. Hunt is an expert in the legal aspects of testing and is active in various industry associations, including the Performance Testing Council where he serves as general counsel.

Stuart Kahl, PhD, is the president, CEO, and cofounder of Measured Progress, Inc., a company that has specialized in customized large-scale educational assessment programs since its inception 24 years ago. Dr. Kahl has more than 30 years of direct experience in the design, development, analysis, and reporting associated with large-scale assessments. He has a BA in mathematics as well as a MEd from Johns Hopkins University and a PhD from the University of Colorado. Prior to entering the testing industry, Dr. Kahl taught graduate courses in statistics/measurement and worked on federally-funded research projects in science and mathematics education. He speaks frequently at meetings and conferences of professional education groups and provides technical consulting to various agencies.

Joan E. Knapp, PhD, is the CEO of Knapp & Associates International, Inc. (K&AI), a firm she founded in 1989 to serve certification organizations, professional and trade associations, and the testing industry. Previously, she was executive director of Credentialing Programs in the Center of Occupational and Professional Assessment at Educational Testing Service. Dr. Knapp has actively participated in the design and implementation of the two main accreditation programs in the certification industry, currently serving on the ANSI Accreditation Committee for Personnel Certifiers and previously serving as psychometrician on the NCCA accreditation program. She is co-chair of the Certification and Licensing Division of ATP and received the NOCA President's award for contributions to NOCA and certification organizations.

Peter Kronvall, MS, heads the certification division of the Swedish national accreditation body SWEDAC (1994–present). Born and raised in Karlstad, Kronvall graduated with an MS in engineering from the Chalmers University of Technology in Gothenburg. His previous positions include six years with Skanska, a global construction services group. Kronvall chaired the IAF task force guidance to ISO/IEC 17024. He also participated in the IAO WG 17 when preparing ISO/IEC 17024.

Dennis Maynes, MA, work interests and current emphasis are in the development and usage of testing models to identify for change and aberrant patterns in item response. He is also actively pursuing applied research in optimal sequential model selection for pattern recognition. He specializes in linear and nonlinear modeling using regression, neural networks, and sequential models. He employed his skills during tenures at Wicat Systems and Wicat Education Institute, Intel, and Fonix, a speech recognition company. Maynes holds a master's degree in statistics from Brigham Young University.

Rohit Ramaswamy, PhD, is vice president of Client Relationships at Oriel Inc., a management consulting firm specializing in helping companies with business transformation through the application of Six Sigma and other process improvement methodologies. He is also the president of Service Design Solutions, a consulting company that helps companies design profitable services. Prior to his consulting work, Dr. Ramaswamy worked at AT&T Bell Labs, where he managed AT&T's internal consulting practice on process improvement for business services. Dr. Ramaswamy taught psychometrics while he was a visiting research scholar at the Department of Educational Psychology at Rutgers University, and has consulted on process improvements in the testing industry. Dr. Ramaswamy is the author of *Design and Management of Service Processes* (Prentice-Hall, 1996), the only currently available book on systematic process design. He holds a bachelor's degree in mechanical engineering from the Indian Institute of Technology, and an MS and PhD in transportation systems from the Massachusetts Institute of Technology.

Marten Roorda, MA, is chief executive officer of Cito, Europe's leading testing and assessment company, based in Arnhem, The Netherlands. Cito has foreign branches, participates in innovative joint ventures, and performs projects for international organizations: World Bank, European Union, and OECD (PISA project). Previously, Roorda held management positions at Reed Elsevier, after having an editorial career. Roorda is a member of the Board of Directors of the Association of Test Publishers, the founder of its European Division, and cofounder of Kryterion, an online testing company. He was a keynote speaker during the Association of Test Publishers Annual Conference in 2004.

Michael Rosenfeld, PhD, is president of Rosenfeld & Associates. He has worked in the areas of licensing and certification for more than 20 years. He was the director of Professional and Occupational Studies at the Educational Testing Service for many years. He is coeditor of *The CLEAR Exam Review,* a journal dealing with issues of licensing and certification, and is the lead assessor for the American National Standards Institute in its program that accredits credentialing organizations. Dr. Rosenfeld has authored or coauthored more than 100 articles and technical reports and has presented numerous papers at professional meetings. He holds a PhD in industrial psychology from Purdue University.

Richard C. Smith is an executive consultant and advisor to PricewaterhouseCoopers (PwC) Senior Partners. Prior to this engagement, he was a former partner in the Strategic Change Practice and the Partner-in-Charge of the Center of Excellence for Six Sigma Process Improvement with IBM and PwC Consulting. He has over 15 years of strategy, change management, and process consulting experience, including consulting in various organizations in the testing industry. He also has more than 18 years of corporate background in sales and marketing, operations, and quality management. Mr. Smith is a speaker and author on the topics of strategy deployment, organizational change, process reengineering, and Six Sigma deployment. He is co-author of the book *Strategic Six Sigma: Best Practices from the Executive Suite,* a book that features the best practices of companies that have successfully deployed Six Sigma as an organizational business initiative.

Judson (Jud) Turner, JD, presently serves in the administration of Georgia Governor Sonny Perdue as Perdue's executive counsel. In addition to legislative work, Turner also advises the governor on a wide array of executive branch legal matters, including government procurement, education policy, and complex litigation. Prior to his work for the governor, Turner served as general counsel at the Georgia Department of Education from 2003 to 2005 and was engaged in a general commercial litigation practice for Bradley, Arant, Rose & White LLP in Birmingham, Alabama, from 2000 to 2003. Turner received his law degree from the University of Virginia and his undergraduate degree from the University of Georgia, where he served as president of the Student Government Association, graduating *Phi Beta Kappa* with highest honors in political science and economics.

Cheryl L. Wild, PhD, is president of Wild & Associates, Inc., a consulting firm that helps its clients to improve their effectiveness and efficiency. Prior to that, she held executive director positions in Test Development and Organizational Development at ETS. She is a lead auditor for two American National Standards Institute accreditation programs, ISO/IEC 17024 and the Conference for Food Protection. Dr. Wild teaches graduate classes in benchmarking at the National Graduate School in Falmouth, Massachusetts. The New Jersey chapter of the Association for Quality and Participation has honored her for directing the New Jersey AQP Team Excellence Competition and for coaching the two Merrill Lynch teams that won the national gold and bronze awards in 2002. Active in the testing industry for more than 25 years, she writes and presents on topics related to testing and quality. She received her doctor of philosophy in educational measurement from Purdue University.

Author Index

Subject Index